Erwin Dee Kord (Ed.)

Semotilus

Erwin Dee Kord (Ed.)

Semotilus

Cyprinidae, Cypriniformes

Solv

Contents

References

Semotilus

Semotilus	
Scientific classification	
Kingdom:	Animalia
Phylum:	Chordata
Class:	Actinopterygii
Order:	Cypriniformes
Family:	Cyprinidae
Genus:	**Semotilus** Rafinesque, 1820

Semotilus is the genus of **creek chubs**, ray-finned fish in the Cyprinidae family. The term "creek chub" is sometimes used for individual species, particularly the Common Creek Chub, *S. atromaculatus*. The Creek Chub are a species of minnows that can grow from 6 to 10 inches and resemble the Grass Carp. They can be found in the United States and Canada in any small stream or creek. They like to hide under small rocks for protection.

Species

- *Semotilus atromaculatus* (Mitchill, 1818) (Common creek chub)
- *Semotilus corporalis* (Mitchill, 1817) (Fallfish)
- *Semotilus lumbee* Snelson & Suttkus, 1978 (Sandhills chub)
- *Semotilus thoreauianus* D. S. Jordan, 1877 (Dixie chub)

Fishing for Creek Chub

Creek Chub will go after almost anything, including powerbait, worms, corn, etc. Surprisingly, they are a very strong fish.

References

- Froese, Rainer, and Daniel Pauly, eds. (2011). Species of *Semotilus* [1] in FishBase. October 2011 version.

Cyprinidae

<table>
<tr><td colspan="2" align="center">Cyprinids</td></tr>
<tr><td colspan="2" align="center"></td></tr>
<tr><td colspan="2" align="center">The Tench (Tinca tinca) is of unclear affiliations and often placed in a subfamily of its own</td></tr>
<tr><td colspan="2" align="center">Scientific classification</td></tr>
<tr><td>Kingdom:</td><td>Animalia</td></tr>
<tr><td>Phylum:</td><td>Chordata</td></tr>
<tr><td>Superclass:</td><td>Osteichthyes</td></tr>
<tr><td>Class:</td><td>Actinopterygii</td></tr>
<tr><td>Subclass:</td><td>Neopterygii</td></tr>
<tr><td>Infraclass:</td><td>Teleostei</td></tr>
<tr><td>Superorder:</td><td>Ostariophysi</td></tr>
<tr><td>Order:</td><td>Cypriniformes</td></tr>
<tr><td>Superfamily:</td><td>Cyprinioidea</td></tr>
<tr><td>Family:</td><td>Cyprinidae</td></tr>
<tr><td colspan="2" align="center">Subfamilies</td></tr>
<tr><td colspan="2">Acheilognathinae
Cultrinae
Cyprininae
Danioninae
Gobioninae
Hypophthalmichthyinae
Labeoninae (disputed)
Leuciscinae
Psilorhynchinae
Rasborinae (polyphyletic?)
Squaliobarbinae (disputed)
Tincinae
and see text</td></tr>
</table>

The family **Cyprinidae**, from the Ancient Greek *kyprînos* (κυπρῖνος, "carp"), consists of the **carps**, the true **minnows**, and their relatives (for example, the **barbs** and **barbels**). Commonly called the **carp family** or the **minnow family**, its members are also known as **cyprinids**. It is the largest family of fresh-water fish, with over 2,400 species in about 220 genera. The family belongs to the order Cypriniformes, of whose genera and species the cyprinids make up two-thirds.[1]

Description

Giant Barbs (*Catlocarpio siamensis*) are the largest members of this family

Cyprinids are stomachless fish with toothless jaws. Even so, food can be effectively chewed by the gill rakers of the specialized last gill bow. These pharyngeal teeth allow the fish to make chewing motions against a chewing plate formed by a procession of the skull. The pharyngheal teeth are species specific and are used by specialists to determine the species. Strong pharyncheal teeth allow fish like the common carp and ide to eat hard baits like snails and bivalves.

Hearing is a well-developed sense, since the cyprinds have the Weberian organ, three specialized vertebra processions that transfer motion of the gas bladder to the inner ear. This construction is also used to observe motion of the gas bladder due to atmospheric conditions or depth changes. The cyprinids are physostomes because the pneumatic duct is retained in adult stages and the fish are able to gulp air to fill the gas bladder or they can dispose excess gas to the gut.

The fish in this family are native to North America, Africa, and Eurasia. The largest cyprinid in this family is the Giant Barb (*Catlocarpio siamensis*), which may grow up to 3 metres (9.8 ft). The largest North American species is the Colorado Pikeminnow (*Ptychocheilus lucius*), of which individuals up to 6 feet (1.8 m) long and weighing over 100 pounds (45 kg) have been recorded.

On the other hand, many species are smaller than 5 centimetres (2.0 in). As of 2008, the smallest known freshwater fish is a cypriniform, *Danionella translucida*, reaching 12 millimetres (0.47 in) at the longest.[2] All fish in this family are egg-layers and most do not guard their eggs, however, there are a few species that build nests and/or guard the eggs. The bitterling-like cyprinids (Acheilognathinae) are notable for depositing their eggs in bivalve molluscs, where the young grow up until able to fend for themselves.

Proud angler with 17 kg Mirror Carp (*Cyprinus carpio*)

Most cyprinids feed mainly on invertebrates and vegetation probably due to the lack of teeth and stomach, but some species like the Asp specialize in fish. Many species ide, common rudd will eat small fish however when reaching a certain size. Even small species like the moderlieschen eat larvae of the common frog in artificial circumstances.

Some fishes, such as the grass carp, are specialized in eating vegetation, some, such as the common nase, eat algae from hard surfaces, some, such as the black carp, specialize in snails, and some, such as the silver carp, are specialized filter feeders. For this reason, they are often introduced as a management tool to control various factors in the aquatic environment, such as aquatic vegetation and diseases transmitted by snails.

Relationship with humans

Cyprinids are highly important food fish; they are fished and farmed across Eurasia. In land-locked countries in particular, cyprinids are often the major species of fish eaten because they make the largest part of biomass in most water types except for fast flowing rivers. In non-landlocked countries they are not very much appreciated due to the high number of bones. In Eastern Europe they are often prepared with traditional methods like drying and salting. The prevalence of inexpensive frozen fish products made this less important now than it was in earlier times. Nonetheless, in certain places they remain popular for food as well as recreational fishing, and have been deliberately stocked in ponds and lakes for centuries for this reason.[3]

Cyprinids are popular for angling especially for match fishing (due to their dominance in biomass and numbers) and fishing for common carp because of its size and strength.

Several cyprinids have been introduced to waters outside their natural range to provide food, sport, or biological control for some pest species. The Common Carp (*Cyprinus carpio*) and the Grass Carp (*Ctenopharyngodon idella*) are the most important of these, for example in Florida. In some cases, these have become invasive species that compete with native fishes or disrupt the environment. Carp in particular can stir up sediment, reducing the clarity of the water and making it difficult for plants to grow.[4]

Numerous cyprinids have become important in the aquarium hobby, most famously the Goldfish, which was bred in China from the Prussian Carp (*Carassius (auratus) gibelio*). First imported into Europe around 1728, it was much fancied by Chinese nobility as early as 1150 AD and after it arrived there in 1502, also in Japan. In the latter country, from the 18th century onwards the Common Carp was bred into the ornamental variety known as koi – or more accurately *nishikigoi* (), as *koi* () simply means "Common Carp" in Japanese.

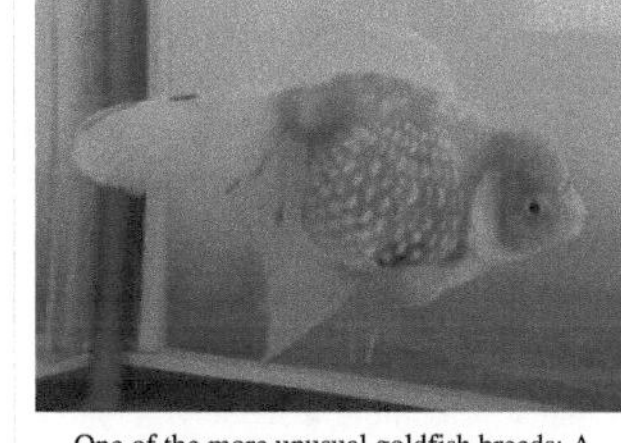
One of the more unusual goldfish breeds: A variegated pearlscale with an oranda-type *wen* ("cap").

Other popular aquarium cyprinids include danionins, rasborines and true barbs.[5] Larger species are bred by the thousands in outdoor ponds, particularly in Southeast Asia, and trade in these aquarium fishes is of considerable commercial importance. The small rasborines and danionines are perhaps only rivalled by characids and poecilid livebearers in their popularity for community aquaria.

One particular species of these small and undemanding danionines is the Zebrafish (*Danio rerio*). It has become the standard model species for studying developmental genetics of vertebrates, in particular fish.[6]

Habitat destruction and other causes have reduced the wild stocks of several cyprinids to dangerously low levels; some are already entirely extinct. In particular, Leuciscinae from southwestern North America have been hit hard by pollution and unsustainable water use in the early-mid 20th century; most globally extinct Cypriniformes species are in fact Leuciscinae from the southwestern United States and northern Mexico.

Systematics

The massive diversity of cyprinids has so far made it difficult to resolve their phylogeny in sufficient detail to make assignment to subfamilies more than tentative in many cases. It is obvious that some distinct lineages exist – for example, Cultrinae and Leuciscinae, regardless of their exact delimitation, are rather close relatives and stand apart from Cyprininae –, but the overall systematics and taxonomy of the Cyprinidae remain a subject of considerable debate. A large number of genera are *incertae sedis*, too equivocal in their traits and/or too little-studied to permit assignment to a particular subfamily with any certainty.[7]

Part of the solution seems that the delicate rasborines are the core group, consisting of minor lineages that have not shifted far from their evolutionary niche, or have co-evolved, for millions of years. These are among the most basal lineages of living cyprinids. Other "rasborines" are apparently distributed across the diverse lineages of the family.[8]

The validity and circumscription of proposed subfamilies like Labeoninae or Squaliobarbinae also remains doubtful, although the latter do appear to correspond to a distinct lineage. The sometimes-seen grouping of the large-headed carps (Hypophthalmichthyinae) with *Xenocypris*, on the other hand, seems quite in error. More likely, the latter are part of the Cultrinae.[8]

The entirely paraphyletic "Barbinae" and the disputed Labeoninae might be better treated as part of the Cyprininae, forming a close-knit group whose internal relationships are still little known. The small African "barbs" do not belong in *Barbus* sensu stricto – indeed, they are as distant from the typical barbels and the typical carps (*Cyprinus*) as these are from *Garra* (which is placed in the Labeoninae by most who accept the latter as distinct) and thus might form another as of yet unnamed subfamily. However, as noted above, how various minor lineages tie into this has not yet been resolved; therefore such a radical move, though reasonable, is probably premature.[9]

Itasenpara Bitterling (*Acheilognathus longipinnis*: Acheilognathinae)

The Tench (*Tinca tinca*), a significant food species farmed in western Eurasia in large numbers, is unusual. It is most often grouped with the Leuciscinae, but even when these were rather loosely circumscribed, it always stood apart. A cladistic analysis of DNA sequence data of the S7 ribosomal protein intron 1 supports the view that it is distinct enough to constitute a monotypic subfamily. It also suggests that it may be closer to the small East Asian *Aphyocypris*, *Hemigrammocypris*, and *Yaoshanicus*. They would have diverged roughly at the same time from cyprinids of east-central Asia, perhaps as a result of the Alpide orogeny that vastly changed the topography of that region in the late Paleogene, when their divergence presumably occurred.[10]

Blue Danio (*Danio kerri*: Danioninae)

Subfamilies & Genera

Subfamily Acheilognathinae – bitterling-like cyprinids

- *Acanthorhodeus* (Spiny bitterlings)
- *Acheilognathus* (Bitterlings)
- *Rhodeus* (Bitterlings)
- *Tanakia* (Bitterlings)

Subfamily Barbinae (Barbs)

- *Acrossocheilus*
- *Balantiocheilos*
- *Barbus* (typical barbels and barbs)
- *Carasobarbus*
- *Clypeobarbus*
- *Diptychus* – "snowtrouts"
- *Luciobarbus*
- *Mesopotamichthys*
- *Oreichthys*
- *Ospatulus*
- *Pseudobarbus* – redfins
- *Puntius* – spotted barbs
- *Schizothorax* – marinkas, "snowtrouts"
- *Sinocyclocheilus* – golden-line fish
- *Spratellicypris*

Pseudogobio esocinus (Gobioninae)

Silver Carp (*Hypophthalmichthys molitrix*: Hypophthalmichthyinae)

Subfamily Cultrinae

- *Anabarilius*
- *Chanodichthys*
- *Culter*
- *Cultrichthys*
- *Hemiculter* (sharpbellies)
- *Ischikauia*
- *Megalobrama*
- *Parabramis* – White Amur Bream
- *Sinibrama*
- *Toxabramis*

Subfamily Cyprininae (true carps)

- *Carassioides*
- *Carassius* (Crucian carps and goldfish)
- *Cyprinus* – typical carps

Subfamily Danioninae – danionins

- *Amblypharyngodon* (carplets)
- *Barilius*
- *Betadevario*
- *Boraras* (rasboras)
- *Brevibora* (rasboras)
- *Chela*
- *Danio* (danios)
- *Danionella*
- *Devario*
- *Esomus* (flying barbs)
- *Horadandia*
- *Inlecypris*
- *Kottelatia*
- *Laubuca*
- *Leptocypris*
- *Luciosoma*
- *Malayochela*
- *Mesobola*
- *Microdevario*
- *Microrasbora*
- *Nematabramis*
- *Neobola*
- *Opsaridium*
- *Opsarius*
- *Paedocypris*
- *Pectenocypris*
- *Raiamas*
- *Rasboroides*
- *Salmophasia* – razorbelly minnows
- *Securicula*

Rohu (*Labeo rohita*) of the disputed Labeoninae)

- *Sundadanio*
- *Trigonostigma*

Subfamily Gobioninae – true gudgeons and relatives (including Gobiobotinae)

- *Coreius*
- *Gnathopogon*
- *Gobio* – typical gudgeons
- *Gobiobotia*
- *Gobiocypris*
- *Hemibarbus* (steeds)
- *Microphysogobio*
- *Pseudogobio*
- *Pseudorasbora*
- *Romanogobio*
- *Sarcocheilichthys*
- *Saurogobio*
- *Squalidus*

Subfamily Labeoninae (including Garrinae; might belong in Cyprininae)

- *Akrokolioplax*
- *Bangana*
- *Cirrhinus* (mud carps)
- *Cophecheilus*
- *Crossocheilus*
- *Discocheilus*
- *Discogobio*
- *Garra*
- *Hongshuia*
- *Labeo* – Labeos
- *Labeobarbus* – yellowfish
- *Labiobarbus*
- *Osteochilus*
- *Parasinilabeo*
- *Protolabeo*
- *Pseudocrossocheilus*
- *Pseudogyrinocheilus*
- *Ptychidio*
- *Qianlabeo*
- *Rectoris*
- *Semilabeo*
- *Sinocrossocheilus*

Subfamily Leptobarbinae

- *Leptobarbus*

Subfamily Leuciscinae – chubs, daces, true minnows, roaches, shiners and so on.

Flame Chub (*Hemitremia flammea*), one of the chubs in the Leuciscinae)

Ide (*Leuciscus idus*), one of the Eurasian daces

Sailfin Shiner (*Notropis hypselopterus*), a small and colorful shiner of the Leuciscinae)

Chinese Minnow (*Rhynchocypris oxycephalus*), a minnow related to some North American daces

Rutilus rubilio a European roach

- *Achondrostoma*
- *Acrocheilus* (Chiselmouth)
- *Agosia*
- *Alburnus* (bleaks)
- *Algansea* (chubs)
- *Aztecula* (Aztec shiner)
- *Campostoma* (stonerollers)
- *Chondrostoma* (typical nases)
- *Chrosomus* (daces)
- *Clinostomus* (redside daces)
- *Codoma*
- *Couesius* (the Lake Chub)
- *Cyprinella* – satinfin shiners
- *Delminichthys*
- *Dionda* – desert minnows
- *Eremichthys* – Desert Dace
- *Ericymba* – the Longjaw Minnow
- *Erimonax*
- *Erimystax* – slender chubs
- †*Evarra* – Mexican daces
- *Exoglossum* – cutlips minnows
- *Gila* – western chubs (including *Siphateles*)
- *Hemitremia* – Flame Chub
- *Hesperoleucus* – California Roach
- *Hybognathus* – silvery minnows
- *Hybopsis* – bigeye chubs
- *Iberochondrostoma*
- *Iotichthys* – Least Chub
- *Lavinia* – Hitch
- *Lepidomeda* – spinedaces
- *Leuciscus* – Eurasian daces
- *Luxilus* – highscale shiners
- *Lythrurus* – finescale shiners
- *Macrhybopsis* – blacktail chubs
- *Margariscus* – Daces
- *Meda* – Spikedace
- *Moapa* – Moapa Dace
- *Mylocheilus* – peamouth
- *Mylopharodon* – hardheads
- *Nocomis* – hornyhead chubs
- *Notemigonus* – Golden Shiner
- *Notropis* – eastern shiners
- *Opsopoeodus* – pugnose minnow
- *Oregonichthys* – Oregon chubs
- *Orthodon* – Sacramento Blackfish
- *Parachondrostoma*
- *Pelasgus*
- *Pelecus* – Ziege, Sabre Carp
- *Petroleuciscus* – Ponto-Caspian chubs and daces
- *Phenacobius* – suckermouth minnows
- *Phoxinellus*
- *Phoxinus* – Eurasian minnows and daces
- *Pimephales* – bluntnose minnows
- *Plagopterus* – Woundfin
- *Platygobio* – flathead chub
- *Pogonichthys* – splittails
- *Protochondrostoma*
- *Pseudochondrostoma*
- *Pseudophoxinus*
- *Pteronotropis* – flagfin shiners
- *Ptychocheilus* – pikeminnows
- *Relictus* – Relict Dace
- *Rhinichthys* – riffle daces (including *Tiaroga*)
- *Rhynchocypris* – Eurasian minnows
- *Richardsonius* – redside shiners
- *Semotilus* – creek chubs
- *Snyderichthys* – Leatherside Chub
- *Squalius* – European chubs
- *Telestes*
- *Tribolodon*
- *Yuriria*

Trigonostigma somphongsi ("Rasborinae", probably not too distant from the Blue Danio above)

Black Carp (*Mylopharyngodon piceus*: Squaliobarbinae)

Subfamily Rasborinae – rasborines

- *Aphyocypris*
- *Aspidoparia*
- *Engraulicypris* (the Lake Sardine)
- *Oxygaster*
- *Rasbora*
- *Rasbosoma*
- *Rastrineobola* – Silver Cyprinid
- *Thryssocypris*
- *Trigonopoma*

Subfamily Squaliobarbinae

- *Ctenopharyngodon* – the Grass Carp
- *Squaliobarbus*

Subfamily Tincinae

- *Tanichthys* – cardinal minnows
- *Tinca* – Tench

Subfamily Xenocyprinae

- *Distoechodon*
- *Hypophthalmichthys* – bighead carps
- *Plagiognathops*
- *Pseudobrama*
- *Xenocypris*

Incertae sedis

Hemigrammocypris rasborella is of uncertain relationships.
It might be close to *Aphyocypris*.

- *Aaptosyax* (Giant salmon carp)
- *Abbottina* (False gudgeons)
- *Abramis* (Common bream)
- *Acanthalburnus* (Bleaks)
- *Acanthobrama* (Bleaks)
- *Acanthogobio*
- *Acapoeta*
- *Albulichthys*
- *Alburnoides* (Bleaks)
- *Amblyrhynchichthys*
- *Anaecypris*
- *Ancherythroculter*
- *Anchicyclocheilus*
- *Araiocypris*
- *Aspiolucius*
- *Aspiorhynchus*
- *Aspius*
- *Atrilinea*
- *Aulopyge* (Dalmatian Barbelgudgeon)
- *Ballerus* (Breams)
- *Barbichthys*
- *Barbodes*
- *Barboides*
- *Barbonymus* (Tinfoil barbs)
- *Barbopsis* (the Somalian Blind Barb)
- *Belligobio*
- *Biwia*
- *Blicca* (Silver Bream)
- *Caecobarbus* (the Congo blind barb)
- *Caecocypris*
- *Candidia*
- *Capoeta* (Khramulyas)
- *Capoetobrama*
- *Catla* (the Catla) (Note: some authorities consider this species to belong in the genus *Gibelion*)
- *Catlocarpio*
- *Chagunius*
- *Chelaethiops*
- *Chuanchia*
- *Coptostomabarbus*
- *Coreoleuciscus* (the Korean Splendid Dace)
- *Cosmochilus*
- *Cyclocheilichthys*
- *Cyprinion*
- *Diplocheilichthys*
- *Discherodontus*
- *Discolabeo*
- *Eirmotus*
- *Elopichthys*
- *Epalzeorhynchos*
- *Folifer*
- *Mesogobio*
- *Metzia*
- *Mylopharyngodon* – Black Carp
- *Mystacoleucus*
- *Naziritor* – Zhobi mahseers
- *Neobarynotus*
- *Neolissochilus* – "mahseers"
- *Nicholsicypris*
- *Nipponocypris*
- *Ochetobius*
- *Onychostoma*
- *Opsariichthys*
- *Oreoleuciscus*
- *Osteobrama*
- *Osteochilichthys*
- *Oxygymnocypris*
- *Pachychilon*
- *Paracanthobrama*
- *Parachela*
- *Paracrossochilus*
- *Paralaubuca*
- *Paraleucogobio*
- *Parapsilorhynchus*
- *Pararasbora*
- *Pararhinichthys* – Cheat Minnow
- *Parasikukia*
- *Paraspinibarbus*
- *Parasqualidus*
- *Parator*
- *Parazacco*
- *Percocypris*
- *Phreatichthys* – Somalian Cavefish
- *Placogobio*
- *Platypharodon*
- *Platysmacheilus*
- *Pogobrama*
- *Poropuntius*
- *Probarbus*
- *Procypris*
- *Prolabeo*
- *Prolabeops*
- *Pseudaspius*
- *Pseudobrama*
- *Pseudohemiculter*
- *Pseudolaubuca*
- *Pseudopungtungia*
- *Ptychobarbus*
- *Pungtungia*
- *Puntioplites*
- *Rasborichthys*

* *Gymnocypris*
* *Gymnodanio*
* *Gymnodiptychus*
* *Hainania*
* *Hampala*
* *Hemiculterella*
* *Hemigrammocapoeta*
* *Hemigrammocypris* (close to *Aphyocypris*?)
* *Henicorhynchus*
* *Herzensteinia*
* *Horalabiosa*
* *Huigobio*
* *Hypselobarbus*
* *Hypsibarbus*
* *Iberocypris*
* *Iranocypris* – Iran cave barb
* *Kalimantania*
* *Kosswigobarbus*
* *Ladigesocypris*
* *Ladislavia*
* *Laocypris*
* *Lepidopygopsis*
* *Leucalburnus*
* *Leucaspius* – Moderlieschen
* *Linichthys*
* *Lobocheilos*
* *Longanalus*
* *Longiculter*
* *Luciobrama*
* *Luciocyprinus*
* *Macrochirichthys* – Long Pectoral-fin Minnow
* *Megarasbora*
* *Mekongina*
* *Rhinogobio*
* *Rohtee* – Vatani rohtee
* *Rohteichthys*
* *Rostrogobio*
* *Rutilus* – roaches
* *Sanagia*
* *Sawbwa* – Sawbwa Barb
* *Scaphiodonichthys*
* *Scaphognathops*
* *Scardinius* – rudds
* *Schismatorhynchos*
* *Schizocypris* – "snowtrouts"
* *Schizopyge* – "snowtrouts"
* *Schizopygopsis* – "snowtrouts"
* *Semiplotus*
* *Sikukia*
* *Sinilabeo*
* *Spinibarbus*
* †*Stypodon* – Stumptooth minnow
* *Tampichthys*
* *Thynnichthys*
* *Tor* – "mahseers"
* *Troglocyclocheilus*
* *Tropidophoxinellus*
* *Typhlobarbus*
* *Typhlogarra* – Iraq blind barb
* *Varicorhinus*
* *Vimba*
* *Xenobarbus*
* *Xenocyprioides*
* *Xenophysogobio*
* *Yaoshanicus*
* *Zacco*

Unlike most fish species, cyprinid fish generally increase in abundance in eutrophic lakes. Here, they contribute towards positive feedback as they are efficient at eating zooplankton which would otherwise graze on the algae, reducing its abundance.

See also

* List of fish families

Footnotes

[1] FishBase (2004), Nelson (2006), dictionary.com [2009]
[2] Nelson (2006)
[3] Magri MacMahon (1946): pp.149-152
[4] GSMFC (2005), FFWCC [2008]
[5] Riehl & Baensch (1996): p.410
[6] Helfman *et al.* (1997): p.228
[7] de Graaf *et al.* (2007), He *et al.* (2008a,b)
[8] He *et al.* (2008a)
[9] Howes (1991), de Graaf *et al.* (2007), IUCN (2009)

[10] He *et al.* (2008b)

References

- De Graaf, M.; Megens, H. J.; Samallo, J.; Sibbing, F. A. (2007). "Evolutionary origin of Lake Tana's (Ethiopia) small *Barbus* species: indications of rapid ecological divergence and speciation". *Animal Biology* **57**: 39. doi:10.1163/157075607780002069.
- dictionary.com [2009]: Cyprinid. Retrieved 2009-SEP-25.
- Froese, Rainer, and Daniel Pauly, eds. (2011). "Cyprinidae" (http://www.fishbase.org/Summary/FamilySummary.cfm?Family=Cyprinidae) in FishBase. August 2011 version.
- Florida Fish and Wildlife Conservation Commission (FFWCC) (2006): Florida's Exotic Freshwater Fishes (http://floridafisheries.com/Fishes/non-native.html). Retrieved 2007-03-05.
- Gulf States Marine Fisheries Commission (GSMFC) (2005): Cyprinus carpio (Linnaeus, 1758) (http://nis.gsmfc.org/nis_factsheet2.php?toc_id=183). Version of 2005-08-03. Retrieved 2007-05-03.
- He, S.; Mayden, R.; Wang, X.; Wang, W.; Tang, K.; Chen, W.; Chen, Y. (2008). "Molecular phylogenetics of the family Cyprinidae (Actinopterygii: Cypriniformes) as evidenced by sequence variation in the first intron of S7 ribosomal protein-coding gene: Further evidence from a nuclear gene of the systematic chaos in the family". *Molecular Phylogenetics and Evolution* **46** (3): 818. doi:10.1016/j.ympev.2007.06.001. PMID 18203625.
- He, S.; Gu, X.; Mayden, R. L.; Chen, W. J.; Conway, K. W.; Chen, Y. (2008). "Phylogenetic position of the enigmatic genus Psilorhynchus (Ostariophysi: Cypriniformes): Evidence from the mitochondrial genome". *Molecular Phylogenetics and Evolution* **47** (1): 419. doi:10.1016/j.ympev.2007.10.012. PMID 18053751.
- Helfman, Gene (1997). *The Diversity of Fishes*. Oxford: Blackwell Science. ISBN 0865422567.
- Howes, G.J. (1991): Systematics and biogeography: an overview. *In:* Winfield, I.J. & Nelson, J.S. (eds.): *Biology of Cyprinids*: 1–33. Chapman and Hall Ltd., London.
- International Union for the Conservation of Nature and Natural Resources (IUCN) (2009): *2009 IUCN Red List of Threatened Species* (http://www.iucnredlist.org). Version 2009.1. Retrieved 2009-SEP-20.
- Magri MacMahon, A.F. (1946): *Fishlore*. Pelican Books.
- Nelson, Joseph (2006). *Fishes of the World*. Chichester: John Wiley & Sons. ISBN 0471250317.
- Riehl, Rudiger (1996). *Aquarium Atlas*. Voyageur Press (MN). ISBN 3882440503.

External links

- "Cyprinidae" (http://www.itis.gov/servlet/SingleRpt/SingleRpt?search_topic=TSN&search_value=163342). Integrated Taxonomic Information System. Retrieved 28 April 2004.

Cypriniformes

The **Cypriniformes** are an order of ray-finned fish, including the carps, minnows, loaches and relatives. This order contains 5-6 families,[1] over 320 genera, and more than 3,250 species, with new species being described every few months or so, and new genera being recognized regularly. They are most diverse in southeastern Asia, but are entirely absent from Australia and South America.[2]

Their closest living relatives are the Characiformes (characins and allies), the Gymnotiformes (electric eel and American knifefishes) and the Siluriformes (catfishes).[3]

Description

Like other orders of the Ostariophysi, the cypriniformes possess a Weberian apparatus. However, they differ from most of their relatives in having only a dorsal fin on their back; most other Ostariophysi have a small fleshy adipose fin behind the dorsal fin. Further differences are the Cypriniformes' kinethmoid and the lack of teeth in the mouth. Instead, they have convergent structures called pharyngeal teeth in the throat. While other groups of fish, such as cichlids, also possess pharyngeal teeth, the cypriniformes' teeth grind against a chewing pad on the base of the skull, instead of an upper pharyngeal jaw.[2]

A true loach: the Spined Loach, *Cobitis taenia*

The most notable family placed here is the Cyprinidae (carps and minnows) which make up two-thirds of the order's diversity. This is one of the largest families of fish, and is widely distributed across Africa, Eurasia, and North America. Most species are strictly freshwater inhabitants, but a considerable number are found in brackish water, such as roach and bream. At least one species is found in the sea, the Pacific Redfin, *Tribolodon brandtii*.[4] Brackish water and marine cyprinids are invariably anadromous, swimming upstream into rivers to spawn. The enigmatic mountain carps are a small group of mountain stream fishes confined to Southeast Asia. Sometimes separated as family Psilorhynchidae, they seem to be specially-adapted Cyprinidae.[5]

The Balitoridae and Gyrinocheilidae are families of mountain stream fishes feeding on algae and small invertebrates. They are found only in tropical and subtropical Asia. While the former are a speciose group, the latter contain only a handful of species.[6] The suckers (Catostomidae) are found in temperate North America and eastern Asia. These large fishes are similar to carps in appearance and ecology. The Cobitidae are common across Eurasia and parts of North Africa. A mid-sized group like the suckers,[7] they are rather similar to catfish in appearance and behaviour, feeding primarily off the substrate and equipped with barbels to help them locate food at night or in murky conditions. The Cobitidae, Balitoridae, and Gyrinocheilidae are called loaches, although it seems that the last do not belong to the lineage of "true" loaches but are related to the suckers.[8]

Systematics and evolution

Historically these included all the forms now placed in the superorder Ostariophysi except the catfish, which were placed in the order Siluriformes. By this definition, the Cypriniformes were paraphyletic, so recently the orders Gonorhynchiformes, Characiformes (characins and allies), and Gymnotiformes (knifefishes and electric eels) have been separated out to form their own monophyletic orders.[9]

The families of Cypriniformes are traditionally divided into two superfamilies. Superfamily Cyprinioidea contains the carps and minnows (Cyprinidae) and, according to some, also the mysterious mountain carps as the family Psilorhynchidae. The superfamily Cobitioidea contains hillstream loaches (Balitoridae), suckers (Catostomidae), true loaches (Cobitidae), and sucking loaches (Gyrinocheilidae) in the traditional system.[2]

Nemacheilus chrysolaimos is a hillstream loach. Closely related to true loaches, like these they have barbels.

Catostomoidea is usually treated as a junior synonym of Cobitioidea. But it seems that it could be split off the Catostomidae and Gyrinocheilidae in a distinct superfamily; the Catostomoidea might be closer relatives of the carps and minnows than of the "true" loaches. While the Cyprinioidea seem more "primitive" than the loach-like forms,[2] they were apparently successful enough never to shift from the original ecological niche of the basal Ostariophysi. Yet, from the ecomorphologically conservative main lineage apparently at least two major radiations branched off. These diversified from the lowlands into torrential river habitats, acquiring similar habitus and adaptations in the process.[8]

The Chinese Algae Eater (*Gyrinocheilus aymonieri*), one of the sucking loaches which are distant from other "loaches".

The mountain carps are highly apomorphic Cyprinidae, perhaps close to true carps (Cyprininae), or maybe to the danionins. While some details about the phylogenetic structures of this massively diverse family are known – e.g. that Cultrinae and Leuciscinae are rather close relatives and stand apart from Cyprininae – there is no good consensus yet on how the main lineages are interrelated. A systematic list, from the most ancient to the most modern lineages, can thus be given as:[8]

Erimyzon sucetta, a small sucker.

Superfamily Cobitioidea

- Family Balitoridae – hillstream loaches
- Family Cobitidae – true loaches

Superfamily Catostomoidea

- Family Catostomidae – suckers
- Family Gyrinocheilidae – sucking loaches

Superfamily Cyprinioidea

- Family Cyprinidae – carps (including koi and goldfish) and minnows (including Psilorhynchidae)

Evolution

Cypriniformes include the most primitive of the Ostariophysi in the narrow sense (i.e. excluding Gonorynchiformes). This is evidenced not only by physiological details, but their great distribution, which indicates they had the longest time to spread. The earliest that Cypriniformes might have diverged from Characiphysi (Characiformes and relatives) is thought to be about the Early Triassic, about 250 million years ago (mya).[10] However, their divergence probably occurred only with the splitting-up of Pangaea in the Jurassic, maybe 160 million years ago. By 110 mya, the plate tectonics evidence indicates that the Laurasian Cypriniformes must have been distinct from their Gondwanan relatives.[11]

Cypriniformes is thought to have originated in south-east Asia, where the most diversity of this group is found today. The alternative hypothesis is that they began in South America, similar to the other otophysans. If this were the case, they would have spread to Asia through Africa or North America. As the Characiformes began to diversify and spread, they may have out-competed South American basal cypriniforms in Africa, where more advanced cypriniforms survive and coexist with characiforms.[12]

The earliest fossils are already assignable to the living family Catostomidae; from the Paleocene of Alberta, they are roughly 60 million years old. During the Eocene (55-35 mya), catostomids and cyprinids spread throughout Asia. In the Oligocene, around 30 mya, advanced cyprinids began to out-compete catostomids wherever they were sympatric, causing a decline of the suckers. Cyprinids reached North America and Europe by about the same time, and Africa in the early Miocene (some 23-20 mya). The cypriniforms spread to North America through the Bering land bridge, which formed and disappeared again several times during the many millions of years of cypriniform evolution.[12]

Relationship with humans

The Cyprinidae in particular are important in a variety of ways. Many species are important food fish, particularly in Europe and Asia. Some are also important as aquarium fish, of which the goldfish and koi are perhaps the most celebrated. The other families are of less commercial importance. The Catostomidae have some importance in angling, and some "loaches" are bred for the international aquarium fish trade.

Accidentally or deliberately introduced populations of Common Carp (*Cyprinus carpio*) and Grass Carp (*Ctenopharyngodon idella*) are found on all continents except Antarctica. In some cases, these exotic species have a negative impact on the environment. Carp in particular stir up the riverbed reducing the clarity of the water, making it difficult for plants to grow.[13]

In science, one of the most famous members of the Cypriniformes is the zebrafish (*Danio rerio*). The zebrafish is one of the most important vertebrate model organisms in biological and biochemical sciences, being used in many kinds of experiments. As, during early development, the zebrafish has a nearly transparent body, it is ideal for studying developmental biology. It is also used for the elucidation of biochemical signaling pathways, among others [14] They are also good pets, but can be shy in bright light and crowded tanks.

Threats and extinction

The Thicktail Chub (*Gila crassicauda*) is globally extinct since about 1960.

Habitat destruction, damming of upland rivers, pollution and in some cases overfishing for food or pet trade have driven some Cypriniformes to the brink of extinction or even beyond. In particular, Cyprinidae of southwestern North America have been severely affected; a considerable number went entirely extinct after settlement by Europeans. For example, in 1900 the Thicktail Chub (*Gila crassicauda*) was the most common freshwater fish found in California; 70 years later not a single living individual existed anymore.

The well-known Red-tailed Black Shark (*Epalzeorhynchos bicolor*) from the Mae Klong river of *The Bridge on the River Kwai* fame possibly only survives in captivity. Ironically, while pollution and other forms of overuse by humans have driven it from its native home, it is bred for the aquarium fish trade by the thousands. The Yarqon Bleak (*Acanthobrama telavivensis*) from the Yarqon River had to be rescued into captivity from imminent extinction; new populations have apparently been established again successfully from captive stock. Balitoridae and Cobitidae, meanwhile, contain a very large number of species about which essentially nothing is known except how they look like and where they were first found.[15]

Few if any Red-tailed Black Sharks (*Epalzeorhynchos bicolor*) remain in the wild today.

Globally extinct Cypriniformes species are:[15]

- *Acanthobrama hulensis*
- Gökçe Balığı, *Alburnus akili*
- *Barbus microbarbis*
- Snake River Sucker, *Chasmistes muriei*
- *Chondrostoma scodrense*
- *Cyprinus yilongensis*
- *Evarra bustamantei*
- *Evarra eigenmanni*
- *Evarra tlahuacensis*
- Thicktail Chub, *Gila crassicauda*
- Pahranagat Spinedace, *Lepidomeda altivelis*
- Harelip Sucker, *Moxostoma lacerum*
- Ameca Shiner, *Notropis amecae*
- Durango Shiner, *Notropis aulidion*
- Phantom Shiner, *Notropis orca*
- Salado Shiner, *Notropis saladonis*
- Clear Lake Splittail, *Pogonichthys ciscoides*
- Las Vegas Dace, *Rhinichthys deaconi*
- Stumptooth Minnow, *Stypodon signifer*
- *Telestes ukliva*

Footnotes

[1] FishBase (2005)
[2] Nelson (2006)
[3] Saitoh *et al.* (2003), Briggs (2005)
[4] Orlov & Sa-a {2007]
[5] FishBase (2004d,f), He *et al.* (2008)
[6] FishBase (2004a,e)
[7] FishBase (2004b,c)
[8] He *et al.* (2008)
[9] Helfman *et al.* (1997): pp.228-229
[10] Saitoh *et al.* (2003)
[11] Briggs (2005), Nelson (2006)
[12] Briggs (2005)
[13] GSMFC (2005), FFWCC [2008]
[14] http://www.zfin.org
[15] IUCN (2007)

References

* Briggs, John C. (2005): The biogeography of otophysan fishes (Ostariophysi: Otophysi): a new appraisal. *J. Biogeogr.* **32**(2): 287–294. doi:10.1111/j.1365-2699.2004.01170.x (HTML abstract)
* FishBase (2004a): Family Balitoridae - River loaches (http://filaman.ifm-geomar.de/Summary/FamilySummary.cfm?id=126). Version of 2004-NOV-22. Retrieved 2007-03-05.
* FishBase (2004b): Family Catostomidae - Suckers (http://filaman.ifm-geomar.de/Summary/FamilySummary.cfm?id=125). Version of 2004-NOV-22. Retrieved 2007-03-05.
* FishBase (2004c): Family Cobitidae - Loaches (http://filaman.ifm-geomar.de/Summary/FamilySummary.cfm?id=127). Version of 2004-NOV-22. Retrieved 2007-03-05.
* FishBase (2004d): Family Cyprinidae - Minnows or carps (http://filaman.ifm-geomar.de/Summary/FamilySummary.cfm?id=122). Version of 2004-NOV-22. Retrieved 2007-03-05.
* FishBase (2004e): Family Gyrinocheilidae - Algae eaters (http://filaman.ifm-geomar.de/Summary/FamilySummary.cfm?id=123). Version of 2004-NOV-22. Retrieved 2007-03-05.
* FishBase (2004f): Family Psilorhynchidae - Mountain carps (http://filaman.ifm-geomar.de/Summary/FamilySummary.cfm?id=124). Version of 2004-NOV-22. Retrieved 2007-03-05.
* FishBase (2005): Order Summary for Cypriniformes (http://filaman.ifm-geomar.de/Summary/OrdersSummary.cfm?order=Cypriniformes). Version of 2005-FEB-15. Retrieved 2007-03-05.
* Florida Fish and Wildlife Conservation Commission (FFWCC) (2006): Florida's Exotic Freshwater Fishes (http://floridafisheries.com/Fishes/non-native.html). Retrieved 2007-03-05.
* Gulf States Marine Fisheries Commission (GSMFC) (2005): Cyprinus carpio (Linnaeus, 1758) (http://nis.gsmfc.org/nis_factsheet2.php?toc_id=183). Version of 2005-08-03. Retrieved 2007-05-03.
* He, Shunping; Gub, Xun; Mayden, Richard L.; Chen, Wei-Jen; Conway, Kevin W. & Chen, Yiyu (2008): Phylogenetic position of the enigmatic genus *Psilorhynchus* (Ostariophysi: Cypriniformes): Evidence from the mitochondrial genome. *Mol. Phylogenet. Evol.* **47**: 419–425. doi:10.1016/j.ympev.2007.10.012 (HTML abstract)
* Helfman, G.; Collette, B. & Facey, D. (1997): *The Diversity of Fishes*. Blackwell Publishing. ISBN 0-86542-256-7
* International Union for Conservation of Nature (IUCN) (2007): *[www.iucnredlist.org 2007 IUCN Red List of Threatened Species]*.
* Nelson, Joseph S. (2006): *Fishes of the World*. John Wiley & Sons, Inc. ISBN 0471250317
* Orlov, Alexei & Sa-a, Pascualita [2007]: FishBase - *Tribolodon brandtii* (http://filaman.ifm-geomar.de/Summary/speciesSummary.php?ID=47843&genusname=Tribolodon&speciesname=brandtii). Retrieved 2007-03-05.
* Saitoh, Kenji; Miya, Masaki; Inoue, Jun G.; Ishiguro, Naoya B. & Nishida, Mutsuminame (2003): Mitochondrial Genomics of Ostariophysan Fishes: Perspectives on Phylogeny and Biogeography. *J. Mol. Evol.* **56**(4): 464–472. doi:10.1007/s00239-002-2417-y PMID 12664166 (HTML abstract)

External links

* Cypriniformes Tree of Life (http://bio.slu.edu/mayden/cypriniformes/home.html)

Actinopterygii

The **Actinopterygii** English pronunciation: /ˌæktɪnɒptəˈrɪdʒi.aɪ/ or **ray-finned fishes** constitute a class or sub-class of the bony fishes.

The ray-finned fishes are so called because they possess lepidotrichia or "fin rays", their fins being webs of skin supported by bony or horny spines ("rays"), as opposed to the fleshy, lobed fins that characterize the class Sarcopterygii which also, however, possess lepidotrichia. These actinopterygian fin rays attach directly to the proximal or basal skeletal elements, the radials, which represent the link or connection between these fins and the internal skeleton (e.g., pelvic and pectoral girdles).

In terms of numbers, actinopterygians are the dominant class of vertebrates, comprising nearly 96% of the 25,000 species of fish. They are ubiquitous throughout fresh water and marine environments from the deep sea to the highest mountain streams. Extant species can range in size from *Paedocypris*, at 8 millimetres (0.31 in), to the massive Ocean Sunfish, at 2300 kilograms (5100 lb), and the long-bodied Oarfish, to at least 11 metres (36 ft).

Fossil record

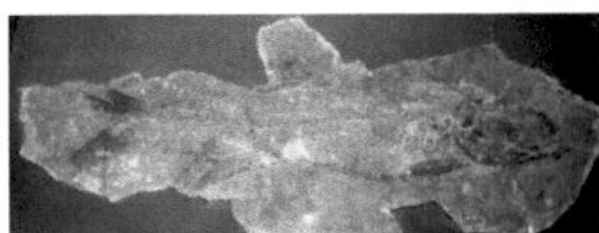

Hypsospondylus fossil

The earliest known fossil Actinopterygiian is *Andreolepis hedei*, dating back 420 million years (Late Silurian). This microvertebrate has been uncovered in Russia, Sweden, and Estonia.[1]

Classification

Traditionally three grades of actinopterygians have been recognised: the **Chondrostei**, **Holostei**, and **Teleostei**. Some morphological evidence suggests that the second is paraphyletic and should be abandoned; however, recent work based on more complete sampling of fossil taxa, and also an analysis of DNA sequence data from the complete mitochondrial genome, supports its recognition. Nearly all living bony fishes are teleosts.

A listing of the different groups is given below, down to the level of orders, arranged in what has been suggested to represent the evolutionary sequence down to the level of order based primarily on the long history of morphological studies. This classification, like any other taxonomy based on phylogenetic research is in a state of flux. Recent morphological and molecular data has shown that several of these ordinal and higher-level groupings represent evolutionary grades rather than clades. Examples of demonstrably paraphyletic groups include the Paracanthopterygii, Scorpaeniformes, and Perciformes.[2] The listing follows FishBase[3] with notes when this differs from Nelson[4] and ITIS.[5]

- **Subclass Chondrostei**
  - Order Polypteriformes, *including the bichirs and reedfishes* [6]
  - Order Acipenseriformes, *including the sturgeons and paddlefishes*
- **Subclass Neopterygii**
  - **Infraclass Holostei**
    - Order Lepisosteiformes, *the gars*
    - Order Amiiformes, *the bowfins*
  - **Infraclass Teleostei**
    - **Superorder Osteoglossomorpha**

- Order Osteoglossiformes, *the bony-tongued fishes*
- Order Hiodontiformes, *including the mooneye and goldeye*
- **Superorder Elopomorpha**
  - Order Elopiformes, *including the ladyfishes and tarpon*
  - Order Albuliformes, *the bonefishes*
  - Order Notacanthiformes, *including the halosaurs and spiny eels*
  - Order Anguilliformes, *the true eels and gulpers*
  - Order Saccopharyngiformes, *including the gulper eel*
- **Superorder Clupeomorpha**
  - Order Clupeiformes, *including herrings and anchovies*
- **Superorder Ostariophysi**
  - Order Gonorynchiformes, *including the milkfishes*
  - Order Cypriniformes, *including barbs, carp, danios, goldfishes, loaches, minnows, rasboras*
  - Order Characiformes, *including characins, pencilfishes, hatchetfishes, piranhas, tetras, dourado / golden (genus Salminus) and pacu.*
  - Order Gymnotiformes, *including electric eels and knifefishes*
  - Order Siluriformes, *the catfishes*
- **Superorder Protacanthopterygii**
  - Order Argentiniformes, *including the barreleyes and slickheads* (formerly in Osmeriformes)
  - Order Salmoniformes, *including salmon and trout*
  - Order Esociformes *the pike*
  - Order Osmeriformes, *including the smelts and galaxiids*
- **Superorder Stenopterygii** (may belong in Protacanthopterygii)
  - Order Ateleopodiformes, *the jellynose fish*
  - Order Stomiiformes, *including the bristlemouths and marine hatchetfishes*
- **Superorder Cyclosquamata** (may belong in Protacanthopterygii)
  - Order Aulopiformes, *including the Bombay duck and lancetfishes*
- **Superorder Scopelomorpha**
  - Order Myctophiformes, *including the lanternfishes*
- **Superorder Lampridiomorpha**
  - Order Lampriformes, *including the oarfish, opah and ribbonfishes*
- **Superorder Polymyxiomorpha**
  - Order Polymixiiformes, *the beardfishes*
- **Superorder Paracanthopterygii**
  - Order Percopsiformes, *including the cavefishes and trout-perches*
  - Order Batrachoidiformes, *the toadfishes*
  - Order Lophiiformes, *including the anglerfishes*
  - Order Gadiformes, *including cods*
  - Order Ophidiiformes, *including the pearlfishes*
- **Superorder Acanthopterygii**
  - Order Mugiliformes, *the mullets*
  - Order Atheriniformes, *including silversides and rainbowfishes*
  - Order Beloniformes, *including the flyingfishes*
  - Order Cetomimiformes, *the whalefishes*
  - Order Cyprinodontiformes, *including livebearers, killifishes*

- Order Stephanoberyciformes, *including the ridgeheads*
- Order Beryciformes, *including the fangtooths and pineconefishes*
- Order Zeiformes, *including the dories*
- Order Gobiesociformes, *the clingfishes*[7]
- Order Gasterosteiformes *including sticklebacks*
- Order Syngnathiformes, *including the seahorses and pipefishes*[8]
- Order Synbranchiformes, *including the swamp eels*
- Order Tetraodontiformes, *including the filefishes and pufferfish*
- Order Pleuronectiformes, *the flatfishes*
- Order Scorpaeniformes, *including scorpionfishes and the sculpins*
- Order Perciformes *40% of all fish including anabantids, centrarchids (incl. bass and sunfish), cichlids, gobies, gouramis, mackerel, tuna, perches, scats, whiting, wrasses*

Notes

[1] Palaeobase (http://paleodb.org/cgi-bin/bridge.pl?action=checkTaxonInfo&taxon_no=34968&is_real_user=1)
[2] G. D. Johnson and E. O. Wiley (March 2007). "Tree of Life: Percomorpha" (http://www.tolweb.org/Percomorpha/52146). .
[3] R. Froese and D. Pauly (editors) (February 2006). "FishBase" (http://www.fishbase.org). .
[4] Nelson, Joseph, S. (2006). *Fishes of the World*. John Wiley & Sons, Inc.. ISBN 0-471-25031-7.
[5] "Actinopterygii" (http://www.itis.gov/servlet/SingleRpt/SingleRpt?search_topic=TSN&search_value=161061). Integrated Taxonomic Information System. . Retrieved 3 April 2006.
[6] In Nelson, Polypteriformes is placed in its own subclass Cladistia.
[7] In ITIS, Gobiesociformes is placed as the suborder Gobiesocoidei of the order Perciformes.
[8] In Nelson and ITIS, Syngnathiformes is placed as the suborder Syngnathoidei of the order Gasterosteiformes.

External links

- *Actinopterygii* (http://www.eol.org/pages/1905) at the Encyclopedia of Life
- Actinopterygii at UntamedScience (http://www.untamedscience.com/biodiversity/animals/chordates/ray-finned-fishes)

Chordate

WARNING: Article could not be rendered - ouputting plain text.

Potential causes of the problem are: (a) a bug in the pdf-writer software (b) problematic Mediawiki markup (c) table is too wide

ChordataTemporal range: Early Cambrian – Recent, X-ray tetra (Pristella maxillaris), one of the few chordates with a visible Vertebral columnbackbone. The spinal cord is housed within its backbone.Biological classificationScientific classification e Kingdom: AnimalAnimalia Superphylum: Deuterostomia Phylum: ChordataWilliam BatesonBateson, 1885Class (biology)Classes See Chordate#ClassificationbelowChordates (phylum Chordata) are animals which are either vertebrates or one of several closely related invertebrates. They are united by having, for at least some period of their life cycle, a notochord, a hollow dorsal nerve cord, pharyngeal slits, an endostyle, and a post-anal tail. The phylum Chordata consists of three subphyla: TunicateTunicata, represented by tunicates; Cephalochordata, represented by lancelets; and Craniata, which includes vertebrateVertebrata. The Hemichordata have been presented as a fourth chordate subphylum, but they are now usually treated as a separate phylum. Tunicate larvae have both a notochord and a nerve cord which are lost in adulthood. Cephalochordates have a notochord and a nerve cord (but no brain or specialist sensory organs) and a very simple circulatory system. Craniates are the only sub-phylum whose members have skulls. In all craniates except for hagfish, the dorsal hollow nerve cord is surrounded with cartilaginous or bony vertebrae and the notochord is generally reduced; hence, hagfish are not regarded as vertebrates. The chordates and three sister phylumphyla, the Hemichordata, the Echinodermata and the Xenoturbellida, make up the deuterostomes, one of the two superphylumsuperphyla that encompass all fairly complex animals.Attempts to work out the evolutionary relationships of the chordates have produced several hypotheses. The current consensus is that chordates are monophyletic, meaning that Chordata contains all and only the descendants of a single common ancestor which is itself a chordate, and that craniates' nearest relatives are cephalochordates. All of the earliest chordate fossils have been found in the Early Cambrian Chengjiang fauna, and include two species that are regarded as fish, which implies that they are vertebrates. Because the fossil record of chordates is poor, only molecular phylogenetics offers a reasonable prospect of dating their emergence. However, the use of molecular phylogenetics for dating evolutionary transitions is controversial.It has also proved difficult to produce a detailed classification within the living chordates. Attempts to produce evolutionary "phylogenetic treefamily trees" give results that differ from traditional class (biology)classes because several of those classes are not monophyletic. As a result vertebrate classification is in a state of flux. DefinitionAnatomy of the cephalochordate Amphioxus. Bolded items are components of all chordates at some point in their lifetime, and distinguish them from other phyla.1 = bulge in spinal cord ("brain")2 = notochord3 = dorsal nerve cord4 = post-anal tail5 = anus6 = alimentary canal digestive canal7 = circulatory system8 = atriopore9 = space above pharynx10 = pharyngeal slit (gill) 11 = pharynx12 = Vestibule of mouthvestibule13 = oral ciliumcirri14 = mouth opening 15 = gonads (ovary / testicle) 16 = light sensor 17 = nerves18 = metapleural fold19 = hepatic caecum (liver-like sack) Anatomy of the cephalochordate Amphioxus. Bolded items are components of all chordates at some point in their lifetime, and distinguish them from other phyla. Chordates form a phylum of creatures that are based on a bilateral body plan,Valentine, J.W. (2004). On the Origin of Phyla. Chicago: University Of Chicago Press. p. 7. ISBN 0226845486."Classifications of organisms in hierarchical systems were in use by the seventeenth and eighteenth centuries. Usually organisms were grouped according to their morphological similarities as perceived by those early workers, and those groups were then grouped according to their similarities, and so on, to form a hierarchy." and is defined by having at some stage in their lives all of the following:Rychel,

A.L., Smith, S.E., Shimamoto, H.T., and Swalla, B.J. (2006). "Evolution and Development of the Chordates: Collagen and Pharyngeal Cartilage". Molecular Biology and Evolution 23 (3): 541–549. doi:10.1093/molbev/msj055. PMID 16280542.A notochord, in other words a fairly stiff rod of cartilage that extends along the inside of the body. Among the vertebrate sub-group of chordates the notochord develops into the Vertebral columnspine, and in wholly aquatic species this helps the animal to swim by flexing its tail. A Anatomical terms of location#Dorsal and ventraldorsal neural tube. In fish and other vertebrates this develops into the spinal cord, the main communications trunk of the nervous system. Pharyngeal slits. The pharynx is the part of the throat immediately behind the mouth. In fish the slits are modified to form gills, but in some other chordates they are part of a filter feedingfilter-feeding system that extracts particles of food from the water in which the animals live. A muscular tail that extends backwards behind the anus. An endostyle. This is a groove in the Anatomical terms of location#Dorsal and ventralventral wall of the pharynx. In filter feedingfilter-feeding species it produces mucus to gather food particles, which helps in transporting food to the esophagus.Ruppert, E. (2005). "Key characters uniting hemichordates and chordates: homologies or homoplasies?". Canadian Journal of Zoology 83: 8–23. doi:10.1139/Z04-158. . Retrieved 2008-09-22. It also stores iodine, and may be a precursor of the vertebrate thyroid gland.Sub-divisionsCraniataCraniate: HagfishCraniates, one of the three sub-divisions of chordates, have distinct skulls - including Hagfish, which have no vertebrae. Michael J. Benton comments that "craniates are characterized by their heads, just as chordates, or possibly all deuterostomes, are by their tails." Benton, M.J. (2000). Vertebrate Palaeontology: Biology and Evolution. Blackwell Publishing. pp. 12–13. ISBN 0632056142. . Retrieved 2008-09-22.Most are vertebrates, in which the notochord is replaced by the spinal column. "Morphology of the Vertebrates". University of California Museum of Paleontology. . Retrieved 2008-09-23.This consists of a series of bony or cartilaginous cylinder (geometry)cylindrical vertebrae, generally with neural arches that protect the spinal cord and with projections that link the vertebrae. Hagfish have incomplete braincases and no vertebrae, and are therefore not regarded as vertebrates, "Introduction to the Myxini". University of California Museum of Paleontology. . Retrieved 2008-10-28. but as members of the craniates, the group from which vertebrates are thought to have evolutionevolved.Neil A. CampbellCampbell, N.A. and Jane ReeceReece, J.B. (2005). Biology (7th ed.). San Francisco, CA: Benjamin Cummings. ISBN 0805370951. The position of lampreys is ambiguous. They have complete braincases and rudimentary vertebrae, and therefore may be regarded as vertebrates and true fish. "Introduction to the Petromyzontiformes". University of California Museum of Paleontology. . Retrieved 2008-10-28. However molecular phylogenetics, which uses biochemistrybiochemical features to classify organisms, has produced both results that group them with vertebrates and others that group them with hagfish.Shigehiro Kuraku, S., Hoshiyama, D., Katoh, K., Suga, H, and Miyata, T. (December 1999). "Monophyly of Lampreys and Hagfishes Supported by Nuclear DNA-Coded Genes". Journal of Molecular Evolution 49 (6): 729–735. doi:10.1007/PL00006595.   PMID 10594174.Cephalochordata: "The   Lancelets"Cephalochordate: LanceletCephalochordates are small, "vaguely fish-shaped" animals that lack brains, clearly defined heads and specialized sense organs.Benton, M.J. (2000). Vertebrate Palaeontology: Biology and Evolution. Blackwell Publishing. p. 6. ISBN 0632056142. . Retrieved 2008-09-22. These burrowing filter-feeders may be either the closest living relatives of craniates or surviving members of the group from which all other chordates evolved.Gee, H. (June 2008). "Evolutionary biology: The amphioxus unleashed". Nature 453 (7198): 999–1000. Bibcode 2008Natur.453..999G. doi:10.1038/453999a. PMID 18563145. . Retrieved 2008-09-22. "Branchiostoma". Lander University. . Retrieved 2008-09-23.Urochordata: "The Tunicates"Tunicates: sea squirtsMost tunicates appear as adults in two major forms, both of which are soft-bodied filter-feeders that lack the standard features of chordates: "sea squirts" are sessile and consist mainly of water pumps and filter-feeding apparatus;Benton, M.J. (2000). Vertebrate Palaeontology: Biology and Evolution. Blackwell Publishing. p. 5. ISBN 0632056142. . Retrieved 2008-09-22. salps float in mid-water, feeding on plankton, and have a two-generation cycle in which one generation is solitary and the next forms chain-like Colony (biology)colonies. "Animal fact files: salp". BBC. . Retrieved 2008-09-22. However all tunicate larvae have the standard chordate features, including long, tadpole-like tails; they also have rudimentary brains, light sensors and tilt sensors. The third main group of tunicates, Appendicularia (also known as Larvacea) retain tadpole-like shapes and active swimming all their lives, and were for a long time regarded as larvae of sea squirts or salps. "Appendicularia" (PDF). Australian Government Department of the Environment, Water, Heritage and the Arts. . Retrieved 2008-10-28. Because of their larvae's long tails tunicates are also called urochordates ("tail chordates").Closest non-chordate relativesThe HemichordatesEnteropneust hemichordate: BakiljkjanoglossusHemichordates ("half(1\2) chordates") have some features similar to those of chordates: branchial openings that open into the pharynx and look rather like gill slits; stomochords, similar in composition to notochords but running in a circle round the "collar", which is ahead of the mouth; and a Anatomical terms of location#Dorsal and ventraldorsal nerve cord — but also a smaller Anatomical terms of location#Dorsal and ventralventral nerve cord. There are two living groups of hemichordates. The solitary enteropneusts, commonly known as "acorn worms", have long proboscisprobosces and worm-like bodies with up to 200 branchial slits, are up to 2.5 metres (8.2 ft) long, and burrow though seafloor sediments. Pterobranchs are colony (biology)colonial animals, often less than 1 millimetre (0.039 in) long individually, whose dwellings are inter-connected. Each filter feedingfilter feeds by means of a pair of branched tentacles, and has a short, shield-shaped proboscis. The extinct graptolites, colonial animals whose fossils look like tiny hacksaw blades, lived in tubes similar to those of pterobranchs. "Introduction to the Hemichordata". University of California Museum of Paleontology. . Retrieved 2008-09-22.The EchinodermsEchinoderm: starfishEchinoderms differ from chordates and their other relatives in three conspicuous ways: instead of having bilateral symmetry they have radial symmetry, meaning their body pattern is shaped like a wheel; they have tube feet; and their bodies are supported by skeletons made of calcite, a material not used by chordates. The hard calcified shell keeps their bodies well protected from the environment, and these skeletons enclose their bodies but are also covered by a thin skin. The feet are powered by another unique feature of echinoderms, a water vascular system of canals that also function as a "lung" and are surrounded by muscles that act as pumps. Crinoids look rather like flowers, and use their feather-like arms to filter food particles out of the water; most live anchored to rocks, but a few can move very slowly. Other echinoderms are mobile and take a variety of body shapes, for example starfish, sea urchins and Holothuroideasea cucumbers.Cowen, R. (2000). History of Life (3rd ed.). Blackwell Science. p. 412. ISBN 063204444-6.Origins The majority of animals more complex than jellyfish and other Cnidarians are split into two groups, the protostomes and deuterostomes, and chordates are deuterostomes.Erwin, Douglas H.; Eric H. Davidson (July 1, 2002). "The last common bilaterian ancestor". Development 129 (13): 3021–3032. PMID 12070079. . It seems very likely that 555 million year old Kimberella was a member of the protostomes.New data on Kimberella, the Vendian mollusc-like organism (White sea region, Russia): palaeoecological and evolutionary implications (2007), "Fedonkin, M.A.; Simonetta, A; Ivantsov, A.Y.", in Vickers-Rich, Patricia; Komarower, Patricia, The Rise and Fall of the Ediacaran Biota, Special publications, 286, London: Geological Society, pp. 157–179, doi:10.1144/SP286.12, ISBN 9781862392335, OCLC 156823511 191881597Butterfield, N.J. (2006). "Hooking some stem-group "worms": fossil lophotrochozoans in the Burgess Shale". Bioessays 28 (12): 1161–6. doi:10.1002/bies.20507. PMID 17120226. If so, this means that the protostome and deuterostome lineages must have split some time before Kimberella appeared — at least 558 million years ago, and hence well before the start of the Cambrian unknown operator: u'cambrianround' million years ago. The Ediacaran fossil Ernietta, from about 549 to 543 million years ago, may represent a deuterostome animal.Dzik , J. (June 1999). "Organic membranous skeleton of the Precambrian metazoans from Namibia". Geology 27 (6): 519–522. Bibcode 1999Geo....27..519D. doi:10.1130/0091-7613(1999)027<0519:OMSOTP>2.3.CO;2. . Retrieved 2008-09-22. Ernettia is from the Kuibis formation, approximate date given by Waggoner, B. (2003). "The Ediacaran Biotas in Space and Time". Integrative and Comparative Biology 43 (1): 104–113. doi:10.1093/icb/43.1.104. PMID 21680415. . Retrieved 2008-09-22.Haikouichthys, from about 518 million years ago in China, may be the earliest known fish.Shu, D-G., Conway Morris, S., and Han, J (January 2003). "Head and backbone of the Early Cambrian vertebrate Haikouichthys". Nature 421 (6922): 526–529. Bibcode 2003Natur.421..526S. doi:10.1038/nature01264. PMID 12556891. . Retrieved 2008-09-21. Fossils of one major deuterostome group, the echinoderms (whose modern members include starfish, sea urchins and crinoids), are quite common from the start of the Cambrian, 542 million years ago.Bengtson, S. (2004). Early skeletal fossils. In Lipps, J.H., and Waggoner, B.M..

"Neoproterozoic-Cambrian Biological Revolutions" (PDF). Paleontological Society Papers 10: 67–78. . Retrieved 2008-07-18. The Mid Cambrian fossil Rhabdotubus johanssoni has been interpreted as a pterobranch hemichordate.Bengtson, S., and Urbanek, A. (October 2007). "Rhabdotubus, a Middle Cambrian rhabdopleurid hemichordate". Lethaia 19 (4): 293–308. doi:10.1111/j.1502-3931.1986.tb00743.x. . Retrieved 2008-09-23. Opinions differ about whether the Chengjiang fauna fossil Yunnanozoon, from the earlier Cambrian, was a hemichordate or chordate.Shu, D., Zhang, X. and Chen, L. (April 1996). "Reinterpretation of Yunnanozoon as the earliest known hemichordate". Nature 380 (6573): 428–430. Bibcode 1996Natur.380..428S. doi:10.1038/380428a0. . Retrieved 2008-09-23.Chen, J-Y., Hang, D-Y., and Li, C.W. (December 1999). "An early Cambrian craniate-like chordate". Nature 402 (6761): 518–522. Bibcode 1999Natur.402..518C. doi:10.1038/990080. . Retrieved 2008-09-23. Another fossil, Haikouella lanceolata, also from the Chengjiang fauna, is interpreted as a chordate and possibly a craniate, as it shows signs of a heart, arteries, gill filaments, a tail, a neural chord with a brain at the front end, and possibly eyes — although it also had short tentacles round its mouth. Haikouichthys and Myllokunmingia, also from the Chengjiang fauna, are regarded as fish.Shu, D-G., Conway Morris, S., and Zhang, X-L. (November 1999). "Lower Cambrian vertebrates from south China" (PDF). Nature 402 (6757): 42. Bibcode 1999Natur.402...42S. doi:10.1038/46965. . Retrieved 2008-09-23. Pikaia, discovered much earlier but from the Mid Cambrian Burgess Shale, is also regarded as a primitive chordate.Shu, D-G., Conway Morris, S., and Zhang, X-L. (November 1996). "A Pikaia-like chordate from the Lower Cambrian of China". Nature 384 (6605): 157–158. Bibcode 1996Natur.384..157S. doi:10.1038/384157a0. . Retrieved 2008-09-23. On the other hand fossils of early chordates are very rare, since non-vertebrate chordates have no bones or teeth, and only one has been reported for the rest of the Cambrian.Conway Morris, S. (2008). "A Redescription of a Rare Chordate, Metaspriggina walcotti Simonetta and Insom, from the Burgess Shale (Middle Cambrian), British Columbia, Canada". Journal of Paleontology 82 (2): 424–430. doi:10.1666/06-130.1. . Retrieved 2009-04-28.DeuterostomesXenoturbellidaHemichordates Echinoderms ChordatesCephalochordates Tunicates (Urochordates) Craniates A consensus family tree of the chordatesPerseke M, Hankeln T, Weich B, Fritzsch G, Stadler PF, Israelsson O, Bernhard D, Schlegel M. (2007) "The mitochondrial DNA of Xenoturbella bocki: genomic architecture and phylogenetic analysis". Theory Biosci. 126(1):35–42. Available online at The evolutionary relationships between the chordate groups and between chordates as a whole and their closest deuterostome relatives have been debated since 1890. Studies based on anatomical, embryologyembryological, and paleontological data have produced different "family trees". Some closely linked chordates and hemichordates, but that idea is now rejected. Combining such analyses with data from a small set of ribosome RNA genes eliminated some older ideas, but open the possibility that tunicates (urochordates) are "basal deuterostomes", in other words surviving members of the group from which echinoderms, hemichordates and chordates evolved.Winchell, C.J., Sullivan, J., Cameron, C.B., Swalla, B.J., and Mallatt, J. (May 1, 2002). "Evaluating Hypotheses of Deuterostome Phylogeny and Chordate Evolution with New LSU and SSU Ribosomal DNA Data". Molecular Biology and Evolution 19 (5): 762–776. PMID 11961109. . Retrieved 2008-09-23. Most researchers agree that, within the chordates, craniates are most closely related to cephalochordates, but there are also reasons for regarding tunicates (urochordates) as craniates' closest relatives.Blair, J.E., and S. Blair Hedges, S.B. (2005). "Molecular Phylogeny and Divergence Times of Deuterostome Animals". Molecular Biology and Evolution 22 (11): 2275–2284. doi:10.1093/molbev/msi225. PMID 16049193. . Retrieved 2008-09-23. One other phylum, Xenoturbellida, appears to be basal within the deuterostomes, in other words closer to the original deuterostomes than to the chordates, echinoderms and hemichordates.Since chordates have left a poor fossil record, attempts have been made to calculate the key dates in their evolution by molecular phylogenetics techniques, in other words by analysing biochemical differences, mainly in RNA. One such study suggested that deuterostomes arose before 900.0 million years ago and the earliest chordates around 896 million years ago. However molecular estimates of dates often disagree with each other and with the fossil record, and their assumption that the molecular clock runs at a known constant rate has been challenged.Ayala, F.J. (1999). "Molecular clock mirages". BioEssays 21 (1): 71–75. doi:10.1002/(SICI)1521-1878(199901)21:1<71::AID-BIES9>3.0.CO;2-B. PMID 10070256. .Schwartz, J. H. and

Maresca, B. (2006). "Do Molecular Clocks Run at All? A Critique of Molecular Systematics". Biological Theory 1 (4): 357–371. doi:10.1162/biot.2006.1.4.357.Classification Taxonomy The following schema is from the third edition of Vertebrate Palaeontology (Benton)Vertebrate Palaeontology.Benton, M.J. (2004). Vertebrate Palaeontology, Third Edition. Blackwell Publishing, 472 pp. The classification scheme is available online The invertebrate chordate classes are from Fishes of the World. Nelson, J. S. (2006). Fishes of the World (4th ed.). New York: John Wiley and Sons, Inc. pp. 601 pp.. ISBN 0-471-25031-7. While it is structured so as to reflect evolutionary relationships (similar to a cladisticscladogram), it also retains the traditional ranks used in Linnaean taxonomy. Phylum Chordata Subphylum UrochordataTunicata (Urochordata) — (tunicates; 3,000 species) Class Ascidiacea(sea squirts) Class Thaliacea (salps) Class Appendicularia (larvaceans) Class Sorberacea (???) Subphylum Cephalochordata (Acraniata) — (lancelets; 30 species) Class Leptocardii (lancelets) Subphylum Vertebrata (Craniata) (vertebrates — animals with backbones; 57,674 species) Class 'Agnatha' paraphyletic (jawless vertebrates; 100+ species) Subclass MyxiniMyxinoidea (hagfish; 65 species) Subclass Petromyzontida (lampreys) Subclass Conodonta† Subclass Pteraspidomorphi (Paleozoic jawless fish) Order Anaspida Order Thelodonti (Paleozoic jawless fish) Infraphylum Gnathostomata (jawed vertebrates) Class Placodermi (Paleozoic armoured forms) Class Chondrichthyes (cartilaginous fish; 900+ species) Class Acanthodii (Paleozoic "spiny sharks") Class Osteichthyes (bony fish; 30,000+ species) Subclass Actinopterygii (ray-finned fish; about 30,000 species) Subclass Sarcopterygii (lobe-finned fish: 8 species) Superclass Tetrapoda (four-limbed vertebrates; 28,000+ species) Class Amphibia (amphibians; 6,000 species) Class Reptilia (reptiles; 8,225+ species) Subclass Anapsida (extinct "proto-reptiles" and possibly turtles) Subclass Diapsida (majority of reptiles, progenitors of birds) Class BirdAves (birds; 8,800–10,000 species) Class Synapsida (mammal-like "reptiles"; 4,500+ species, progenitors of mammals) Class Mammalia (mammals; 5,800 species)Phylogeny Chordates Cladogram of the Chordate phylum. Lines show probable evolutionary relationships, including extinct taxa, which are denoted with a Dagger (typography)dagger, †. Some are invertebrates. The positions (relationships) of the Lancelet, Tunicate, and Craniata clades are as reportedPutnam, H.; Butts, T.; Ferrier, E.; Furlong, F.; Hellsten, U.; Kawashima, T.; Robinson-Rechavi, M.; Shoguchi, E. et al. (Jun 2008). "The amphioxus genome and the evolution of the chordate karyotype". Nature 453 (7198): 1064–1071. Bibcode 2008Natur.453.1064P. doi:10.1038/nature06967. ISSN 0028-0836. PMID 18563158. in the scientific journal Nature (journal)Nature.

Chordata LanceletCephalochordataAmphioxusTunicateTunicataLarvaceaAppendicularia (formerly Larvacea) ThaliaceaAscidiacea CraniataHagfishMyxini VertebrateVertebrataConodonta† Cephalaspidomorphi† Hyperoartia (Petromyzontida) Pteraspidomorphi† GnathostomataPlacodermi† Chondrichthyes TeleostomiAcanthodii† OsteichthyesActinopterygii Sarcopterygiivoid Tetrapoda AmphibianAmphibia AmnioteAmniota Synapsida voidMammalia SauropsidavoidLepidosauromorpha (lizards, snakes, tuatara, and their mosasaurextinct relatives) Archosauromorpha (Crurotarsicrocodiles, birds, and their ornithodiraextinct relatives) See alsoList of chordate ordersReferences External links Complete details about Phylum Chordata at thebigger.com Chordate on GlobalTwitcher.com Chordate node at Tree Of Life Chordate node at NCBI Taxonomy

Animal

<table>
<tr><td colspan="2" align="center">Animals
Temporal range: Ediacaran – Recent</td></tr>
<tr><td colspan="2" align="center">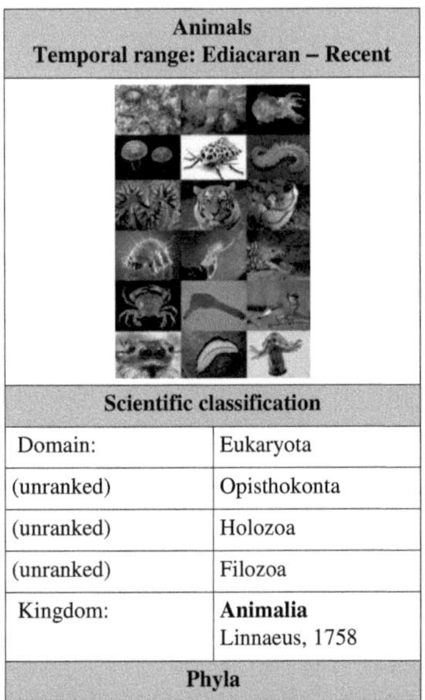</td></tr>
<tr><td colspan="2" align="center">Scientific classification</td></tr>
<tr><td>Domain:</td><td>Eukaryota</td></tr>
<tr><td>(unranked)</td><td>Opisthokonta</td></tr>
<tr><td>(unranked)</td><td>Holozoa</td></tr>
<tr><td>(unranked)</td><td>Filozoa</td></tr>
<tr><td>Kingdom:</td><td>Animalia
Linnaeus, 1758</td></tr>
<tr><td colspan="2" align="center">Phyla</td></tr>
</table>

- **Subkingdom Parazoa**
  - Porifera
  - Placozoa
- **Subkingdom Eumetazoa**
  - **Radiata (unranked)**
    - Ctenophora
    - Cnidaria
  - **Bilateria (unranked)**
    - Orthonectida
    - Rhombozoa
    - Acoelomorpha
    - Chaetognatha
    - **Superphylum Deuterostomia**
      - Chordata
      - Hemichordata
      - Echinodermata
      - Xenoturbellida
      - Vetulicolia †
    - **Protostomia (unranked)**
      - **Superphylum Ecdysozoa**
        - Kinorhyncha
        - Loricifera
        - Priapulida
        - Nematoda
        - Nematomorpha
        - Lobopodia
        - Onychophora
        - Tardigrada
        - Arthropoda
      - **Superphylum Platyzoa**
        - Platyhelminthes
        - Gastrotricha
        - Rotifera
        - Acanthocephala
        - Gnathostomulida
        - Micrognathozoa
        - Cycliophora
      - **Superphylum Lophotrochozoa**
        - Sipuncula
        - Hyolitha †
        - Nemertea
        - Phoronida
        - Bryozoa
        - Entoprocta
        - Brachiopoda
        - Mollusca
        - Annelida
        - Echiura

Animals are a major group of multicellular, eukaryotic organisms of the kingdom **Animalia** or **Metazoa**. Their body plan eventually becomes fixed as they develop, although some undergo a process of metamorphosis later on in their life. Most animals are motile, meaning they can move spontaneously and independently. All animals are also heterotrophs, meaning they must ingest other organisms or their products for sustenance.

Most known animal phyla appeared in the fossil record as marine species during the Cambrian explosion, about 542 million years ago.

Etymology

The word "animal" comes from the Latin word *animalis*, meaning "having breath".[1] In everyday colloquial usage, the word sometimes refers to non-human animals. Sometimes, only closer relatives of humans such as mammals and other vertebrates are meant in colloquial use.[2] The biological definition of the word refers to all members of the kingdom Animalia, encompassing creatures as diverse as sponges, jellyfish, insects and humans.[3]

Characteristics

Animals have several characteristics that set them apart from other living things. Animals are eukaryotic and mostly multicellular,[4] which separates them from bacteria and most protists. They are heterotrophic,[5] generally digesting food in an internal chamber, which separates them from plants and algae.[6] They are also distinguished from plants, algae, and fungi by lacking rigid cell walls.[7] All animals are motile,[8] if only at certain life stages. In most animals, embryos pass through a blastula stage,[9] which is a characteristic exclusive to animals.

Structure

With a few exceptions, most notably the sponges (Phylum Porifera) and Placozoa, animals have bodies differentiated into separate tissues. These include muscles, which are able to contract and control locomotion, and nerve tissues, which send and process signals. Typically, there is also an internal digestive chamber, with one or two openings.[10] Animals with this sort of organization are called metazoans, or eumetazoans when the former is used for animals in general.[11]

All animals have eukaryotic cells, surrounded by a characteristic extracellular matrix composed of collagen and elastic glycoproteins.[12] This may be calcified to form structures like shells, bones, and spicules.[13] During development, it forms a relatively flexible framework[14] upon which cells can move about and be reorganized, making complex structures possible. In contrast, other multicellular organisms, like plants and fungi, have cells held in place by cell walls, and so develop by progressive growth.[10] Also, unique to animal cells are the following intercellular junctions: tight junctions, gap junctions, and desmosomes.[15]

Reproduction and development

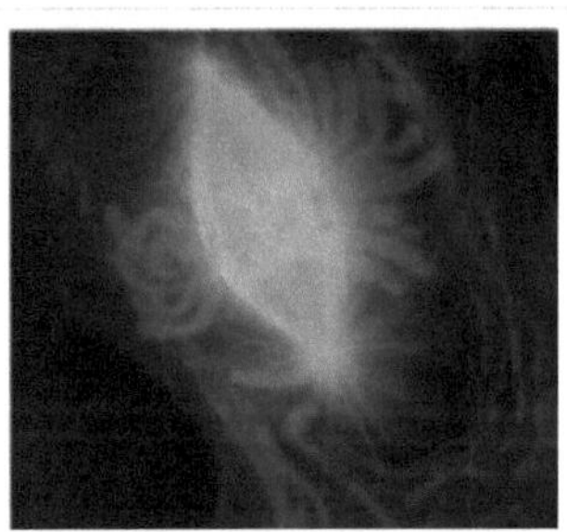
A newt lung cell stained with fluorescent dyes undergoing the early anaphase stage of mitosis

Nearly all animals undergo some form of sexual reproduction.[16] They have a few specialized reproductive cells, which undergo meiosis to produce smaller, motile spermatozoa or larger, non-motile ova.[17] These fuse to form zygotes, which develop into new individuals.[18]

Many animals are also capable of asexual reproduction.[19] This may take place through parthenogenesis, where fertile eggs are produced without mating, budding, or fragmentation.[20]

A zygote initially develops into a hollow sphere, called a blastula,[21] which undergoes rearrangement and differentiation. In sponges, blastula larvae swim to a new location and develop into a new sponge.[22] In most other groups, the blastula undergoes more complicated rearrangement.[23] It first invaginates to form a gastrula with a digestive chamber, and two separate germ layers — an external ectoderm and an internal endoderm.[24] In most cases, a mesoderm also develops between them.[25] These germ layers then differentiate to form tissues and organs.[26]

Food and energy sourcing

All animals are heterotrophs, meaning that they feed directly or indirectly on other living things.[27] They are often further subdivided into groups such as carnivores, herbivores, omnivores, and parasites.[28]

Predation is a biological interaction where a predator (a heterotroph that is hunting) feeds on its prey (the organism that is attacked).[29] Predators may or may not kill their prey prior to feeding on them, but the act of predation always results in the death of the prey.[30] The other main category of consumption is detritivory, the consumption of dead organic matter.[31] It can at times be difficult to separate the two feeding behaviours, for example, where parasitic species prey on a host organism and then lay their eggs on it for their offspring to feed on its decaying corpse. Selective pressures imposed on one another has led to an evolutionary arms race between prey and predator, resulting in various antipredator adaptations.[32]

Most animals indirectly use the energy of sunlight by eating plants or plant-eating animals. Most plants use light to convert inorganic molecules in their environment into organic molecules, such as simple sugars, in photosynthesis. Starting with the molecules carbon dioxide (CO_2) and water (H_2O), photosynthesis converts the energy of sunlight into chemical energy stored as reduced carbon (e.g., glucose) and releases molecular oxygen. These sugars are then used as the building blocks for plant growth.[10] When animals eat these plants (or eat other animals which have eaten plants), the sugars produced by the plant are used by the animal.[33] They are either used directly to help the animal grow, or broken down, releasing stored solar energy, and giving the animal the energy required for motion.[34] This process is known as glycolysis.[35]

Animals living close to hydrothermal vents and cold seeps on the ocean floor are not dependent on the energy of sunlight.[36] Instead chemosynthetic archaea and bacteria form the base of the food chain.[37]

Origin and fossil record

Further information: Urmetazoan

Animals are generally considered to have evolved from a flagellated eukaryote.[39] Their closest known living relatives are the choanoflagellates, collared flagellates that have a morphology similar to the choanocytes of certain sponges.[40] Molecular studies place animals in a supergroup called the opisthokonts, which also include the choanoflagellates, fungi and a few small parasitic protists.[41] The name comes from the posterior location of the flagellum in motile cells, such as most animal spermatozoa, whereas other eukaryotes tend to have anterior flagella.[42]

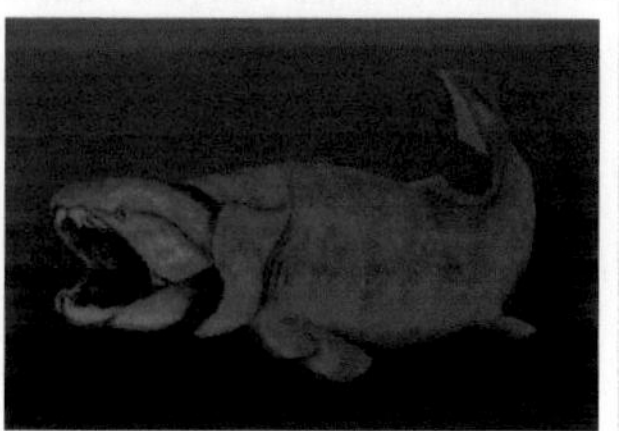

Dunkleosteus was a gigantic, 10-metre-long (33 ft) prehistoric fish.[38]

The first fossils that might represent animals appear in the Trezona Formation at Trezona Bore, West Central Flinders, South Australia.[43] These fossils are interpreted as being early sponges. They were found in 665-million-year-old rock.[43]

The next oldest possible animal fossils are found towards the end of the Precambrian, around 610 million years ago, and are known as the Ediacaran or Vendian biota.[44] These are difficult to relate to later fossils, however. Some may represent precursors of modern phyla, but they may be separate groups, and it is possible they are not really animals at all.[45]

Aside from them, most known animal phyla make a more or less simultaneous appearance during the Cambrian period, about 542 million years ago.[46] It is still disputed whether this event, called the Cambrian explosion, represents a rapid divergence between different groups or a change in conditions that made fossilization possible.

Some paleontologists suggest that animals appeared much earlier than the Cambrian explosion, possibly as early as 1 billion years ago.[47] Trace fossils such as tracks and burrows found in the Tonian era indicate the presence of triploblastic worms, like metazoans, roughly as large (about 5 mm wide) and complex as earthworms.[48] During the beginning of the Tonian period around 1 billion years ago, there was a decrease in Stromatolite diversity, which may indicate the appearance of grazing animals, since stromatolite diversity increased when grazing animals went extinct at the End Permian and End Ordovician extinction events, and decreased shortly after the grazer populations recovered. However the discovery that tracks very similar to these early trace fossils are produced today by the giant single-celled protist *Gromia sphaerica* casts doubt on their interpretation as evidence of early animal evolution.[49] [50]

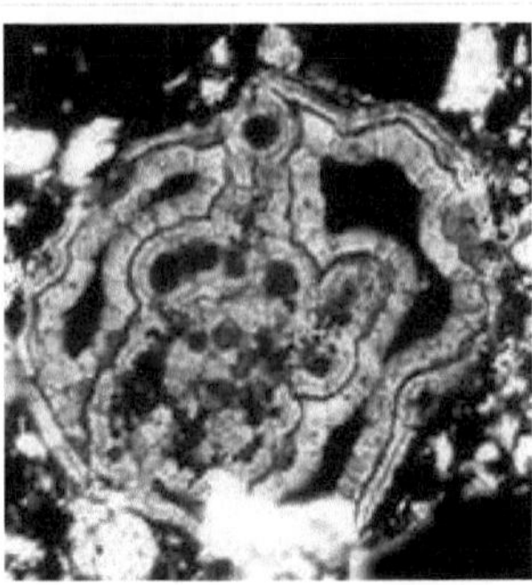

Vernanimalcula guizhouena is a fossil believed by some to represent the earliest known member of the Bilateria.

Groups of animals

Porifera, Radiata and basal Bilateria

Phylogenetic analysis suggests that the Porifera and Ctenophora diverged before a clade that gave rise to the Bilateria, Cnidaria and Placozoa.[51]

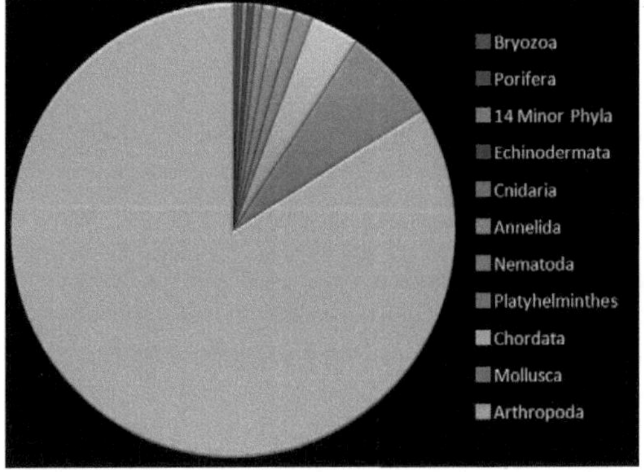

The relative number of species contributed to the total by each phylum of animals.

Orange elephant ear sponge, *Agelas clathrodes*, in foreground. Two corals in the background: a sea fan, *Iciligorgia schrammi*, and a sea rod, *Plexaurella nutans*.

The sponges (Porifera) were long thought to have diverged from other animals early.[52] They lack the complex organization found in most other phyla.[53] Their cells are differentiated, but in most cases not organized into distinct tissues.[54] Sponges typically feed by drawing in water through pores.[55] Archaeocyatha, which have fused skeletons, may represent sponges or a separate phylum.[56] However, a phylogenomic study in 2008 of 150 genes in 29 animals across 21 phyla revealed that it is the Ctenophora or comb jellies which are the basal lineage of animals, at least among those 21 phyla. The authors speculate that sponges—or at least those lines of sponges they investigated—are not so primitive, but may instead be secondarily simplified.[57]

Among the other phyla, the Ctenophora and the Cnidaria, which includes sea anemones, corals, and jellyfish, are radially symmetric and have digestive chambers with a single opening, which serves as both the mouth and the anus.[58] Both have distinct tissues, but they are not organized into organs.[59] There are only two main germ layers, the ectoderm and endoderm, with only scattered cells between them. As such, these animals are sometimes called diploblastic.[60] The tiny placozoans are similar, but they do not have a permanent digestive chamber.

The remaining animals form a monophyletic group called the Bilateria. For the most part, they are bilaterally symmetric, and often have a specialized head with feeding and sensory organs. The body is triploblastic, i.e. all three germ layers are well-developed, and tissues form distinct organs. The digestive chamber has two openings, a mouth and an anus, and there is also an internal body cavity called a coelom or pseudocoelom. There are exceptions to each of these characteristics, however — for instance adult echinoderms are radially symmetric, and certain parasitic worms have extremely simplified body structures.

Genetic studies have considerably changed our understanding of the relationships within the Bilateria. Most appear to belong to two major lineages: the deuterostomes and the protostomes, the latter of which includes the Ecdysozoa, Platyzoa, and Lophotrochozoa. In addition, there are a few small groups of bilaterians with relatively similar structure that appear to have diverged before these major groups. These include the Acoelomorpha, Rhombozoa, and Orthonectida. The Myxozoa, single-celled parasites that were originally considered Protozoa, are now believed to have developed from the Medusozoa as well.

Deuterostomes

Deuterostomes differ from the other Bilateria, called protostomes, in several ways. In both cases there is a complete digestive tract. However, in protostomes, the first opening of the gut to appear in embryological development (the archenteron) develops into the mouth, with the anus forming secondarily. In deuterostomes the anus forms first, with the mouth developing secondarily.[61] In most protostomes, cells simply fill in the interior of the gastrula to form the mesoderm, called schizocoelous development, but in deuterostomes, it forms

Superb Fairy-wren, *Malurus cyaneus*

through invagination of the endoderm, called enterocoelic pouching.[62] Deuterostomes also have a dorsal, rather than a ventral, nerve chord and their embryos undergo different cleavage.

All this suggests the deuterostomes and protostomes are separate, monophyletic lineages. The main phyla of deuterostomes are the Echinodermata and Chordata.[63] The former are radially symmetric and exclusively marine, such as starfish, sea urchins, and sea cucumbers.[64] The latter are dominated by the vertebrates, animals with backbones.[65] These include fish, amphibians, reptiles, birds, and mammals.[66]

In addition to these, the deuterostomes also include the Hemichordata, or acorn worms.[67] Although they are not especially prominent today, the important fossil graptolites may belong to this group.[68]

The Chaetognatha or arrow worms may also be deuterostomes, but more recent studies suggest protostome affinities.

Ecdysozoa

The Ecdysozoa are protostomes, named after the common trait of growth by moulting or ecdysis.[69] The largest animal phylum belongs here, the Arthropoda, including insects, spiders, crabs, and their kin. All these organisms have a body divided into repeating segments, typically with paired appendages. Two smaller phyla, the Onychophora and Tardigrada, are close relatives of the arthropods and share these traits.

Yellow-winged darter, *Sympetrum flaveolum*

The ecdysozoans also include the Nematoda or roundworms, perhaps the second largest animal phylum. Roundworms are typically microscopic, and occur in nearly every environment where there is water.[70] A number are important parasites.[71] Smaller phyla related to them are the Nematomorpha or horsehair worms, and the Kinorhyncha, Priapulida, and Loricifera. These groups have a reduced coelom, called a pseudocoelom.

The remaining two groups of protostomes are sometimes grouped together as the Spiralia, since in both embryos develop with spiral cleavage.

Platyzoa

The Platyzoa include the phylum Platyhelminthes, the flatworms.[72] These were originally considered some of the most primitive Bilateria, but it now appears they developed from more complex ancestors.[73] A number of parasites are included in this group, such as the flukes and tapeworms.[72] Flatworms are acoelomates, lacking a body cavity, as are their closest relatives, the microscopic Gastrotricha.[74]

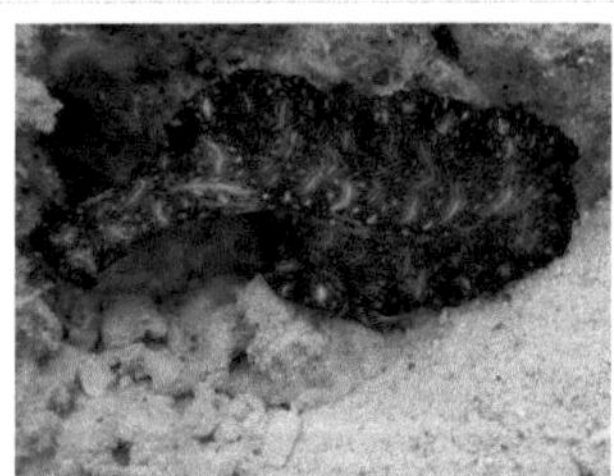

Pseudobiceros bedfordi, (Bedford's flatworm)

The other platyzoan phyla are mostly microscopic and pseudocoelomate. The most prominent are the Rotifera or rotifers, which are common in aqueous environments. They also include the Acanthocephala or spiny-headed worms, the Gnathostomulida, Micrognathozoa, and possibly the Cycliophora.[75] These groups share the presence of complex jaws, from which they are called the Gnathifera.

Lophotrochozoa

The Lophotrochozoa include two of the most successful animal phyla, the Mollusca and Annelida.[76] [77] The former, which is the second-largest animal phylum by number of described species, includes animals such as snails, clams, and squids, and the latter comprises the segmented worms, such as earthworms and leeches. These two groups have long been considered close relatives because of the common presence of trochophore larvae, but the annelids were considered closer to the arthropods because they are both segmented.[78] Now, this is generally considered convergent evolution, owing to many morphological and genetic differences between the two phyla.[79]

Roman snail, *Helix pomatia*

The Lophotrochozoa also include the Nemertea or ribbon worms, the Sipuncula, and several phyla that have a ring of ciliated tentacles around the mouth, called a lophophore.[80] These were traditionally grouped together as the lophophorates.[81] but it now appears that the lophophorate group may be paraphyletic,[82] with some closer to the nemerteans and some to the molluscs and annelids.[83] [84] They include the Brachiopoda or lamp shells, which are prominent in the fossil record, the Entoprocta, the Phoronida, and possibly the Bryozoa or moss animals.[85]

Model organisms

Because of the great diversity found in animals, it is more economical for scientists to study a small number of chosen species so that connections can be drawn from their work and conclusions extrapolated about how animals function in general. Because they are easy to keep and breed, the fruit fly *Drosophila melanogaster* and the nematode *Caenorhabditis elegans* have long been the most intensively studied metazoan model organisms, and were among the first life-forms to be genetically sequenced. This was facilitated by the severely reduced state of their genomes, but as many genes, introns, and linkages lost, these ecdysozoans can teach us little about the origins of animals in general. The extent of this type of evolution within the superphylum will be revealed by the crustacean, annelid, and molluscan genome projects currently in progress. Analysis of the starlet sea anemone genome has emphasised the importance of sponges, placozoans, and choanoflagellates, also being sequenced, in explaining the arrival of 1500 ancestral genes unique to the Eumetazoa.[86]

An analysis of the homoscleromorph sponge *Oscarella carmela* also suggests that the last common ancestor of sponges and the eumetazoan animals was more complex than previously assumed.[87]

Other model organisms belonging to the animal kingdom include the mouse (*Mus musculus*) and zebrafish (*Danio rerio*).

History of classification

Aristotle divided the living world between animals and plants, and this was followed by Carolus Linnaeus (Carl von Linné), in the first hierarchical classification.[88] Since then biologists have begun emphasizing evolutionary relationships, and so these groups have been restricted somewhat. For instance, microscopic protozoa were originally considered animals because they move, but are now treated separately.

In Linnaeus's original scheme, the animals were one of three kingdoms, divided into the classes of Vermes, Insecta, Pisces, Amphibia, Aves, and Mammalia. Since then the last four have all been subsumed into a single phylum, the Chordata, whereas the various other forms have been separated out. The above lists represent our current understanding of the group, though there is some variation from source to source.

Carolus Linnaeus, known as the father of modern taxonomy

See also

- Ethology
- Animal colouration
- Animal rights
- Fauna
- List of animal names
- List of animals by number of neurons
- Lists of animals

References

[1] Cresswell, Julia (2010). *The Oxford Dictionary of Word Origins* (2 ed.). New York: Oxford University Press. ISBN 9780199547937. "'having the breath of life', from anima 'air, breath, life' ."

[2] "Animals" (http://m-w.com/dictionary/animals). Merriam-Webster's. . Retrieved 16 May 2010. "2 a : one of the lower animals as distinguished from human beings b : mammal; *broadly* : vertebrate"

[3] "Animal". *The American Heritage Dictionary* (Forth ed.). Houghton Mifflin Company. 2006.

[4] National Zoo. "Panda Classroom" (http://nationalzoo.si.edu/Animals/GiantPandas/PandasForKids/classification/classification.htm). . Retrieved September 30, 2007.

[5] Jennifer Bergman. "Heterotrophs" (http://www.windows.ucar.edu/tour/link=/earth/Life/heterotrophs.html&edu=high). . Retrieved September 30, 2007.

[6] Douglas AE, Raven JA, AE (2003). "Genomes at the interface between bacteria and organelles". *Philosophical transactions of the Royal Society of London. Series B, Biological sciences* **358** (1429): 5–17; discussion 517–8. doi:10.1098/rstb.2002.1188. PMC 1693093. PMID 12594915.

[7] Davidson, Michael W.. "Animal Cell Structure" (http://micro.magnet.fsu.edu/cells/animalcell.html). . Retrieved September 20, 2007.

[8] Saupe, S.G. "Concepts of Biology" (http://employees.csbsju.edu/SSAUPE/biol116/Zoology/digestion.htm). . Retrieved September 30, 2007.

[9] Minkoff, Eli C. (2008). *Barron's EZ-101 Study Keys Series: Biology* (2, revised ed.). Barron's Educational Series. p. 48. ISBN 9780764139208.

[10] Adam-Carr, Christine; Hayhoe, Christy; Hayhoe, Douglas; Hayhoe, Katharine (2010). *Science Perspectives 10*. Nelson Education Ltd.. ISBN 978-0-17-635528-9.

[11] Gero HIllmer; Ulrich Lehmann (1983). *Fossil Invertebrates*. CUP Archive. p. 54. ISBN 9780521270281.

[12] Alberts, Bruce; Alexander Johnson, Julian Lewis, Martin Raff, Keith Roberts, and Peter Walter (2002). *Molecular Biology of the Cell* (http://www.ncbi.nlm.nih.gov/books/NBK26810/) (4 ed.). New York: Garland Science. .

[13] Sangwal, Keshra (2007). *Additives and crystallization processes: from fundamentals to applications*. John Wiley and Sons. p. 212. ISBN 9780470061534.

[14] Becker, Wayne M. (1991). *The world of the cell*. Benjamin/Cummings. ISBN 9780805308709.

[15] Magloire, Kim (2004). *Cracking the AP Biology Exam, 2004–2005 Edition*. The Princeton Review. p. 45. ISBN 9780375763939.

[16] Knobil, Ernst (1998). *Encyclopedia of reproduction, Volume 1*. Academic Press. p. 315. ISBN 9780122270208.

[17] Schwartz, Jill (2010). *Master the GED 2011 (w/CD)*. Peterson's. p. 371. ISBN 9780768928853.

[18] Hamilton, Matthew B. (2009). *Population genetics*. Wiley-Blackwell. p. 55. ISBN 9781405132770.

[19] Adiyodi, K. G.; Roger N. Hughes, Rita G. Adiyodi (2002). *Reproductive Biology of Invertebrates, Progress in Asexual Reproduction, Volume 11*. Wiley. p. 116.

[20] Kaplan (2008). *GRE exam subject test*. Kaplan Publishing. p. 233. ISBN 9781419552182.

[21] Tmh (2006). *Study Package For Medical College Entrance Examinations*. Tata McGraw-Hill. p. 6.22. ISBN 9780070616370.

[22] Ville, Claude Alvin; Warren Franklin Walker, Robert D. Barnes (1984). *General zoology*. Saunders College Pub. p. 467. ISBN 9780030624513.

[23] Hamilton, William James; James Dixon Boyd, Harland Winfield Mossman (1945). *Human embryology: (prenatal development of form and function)*. Williams & Wilkins. p. 330.

[24] Philips, Joy B. (1975). *Development of vertebrate anatomy*. Mosby. p. 176. ISBN 9780801639272.

[25] *The Encyclopedia Americana: a library of universal knowledge, Volume 10*. Encyclopedia Americana Corp.. 1918. p. 281.

[26] Romoser, William S.; J. G. Stoffolano (1998). *The science of entomology*. WCB McGraw-Hill. p. 156. ISBN 9780697228482.

[27] Rastogi, V. B. (1997). *Modern Biology*. Pitambar Publishing. p. 3. ISBN 9788120904965.

[28] Levy, Charles K. (1973). *Elements of Biology*. Appleton-Century-Crofts. p. 108. ISBN 9780390556271.

[29] Begon, M., Townsend, C., Harper, J. (1996). *Ecology: Individuals, populations and communities* (Third edition). Blackwell Science, London. ISBN 0-86542-845-X, ISBN 0-632-03801-2, ISBN 0-632-04393-8.

[30] predation (http://www.britannica.com/search?query=predation). Britannica.com. Retrieved on 2011-11-23.

[31] Marchetti, Mauro; Victoria Rivas (2001). *Geomorphology and environmental impact assessment*. Taylor & Francis. p. 84. ISBN 9789058093448.

[32] Allen, Larry Glen; Daniel J. Pondella, Michael H. Horn (2006). *Ecology of marine fishes: California and adjacent waters*. University of California Press. p. 428. ISBN 9780520246539.

[33] Clutterbuck, Peter (2000). *Understanding Science: Upper Primary*. Blake Education. p. 9. ISBN 9781865091709.

[34] Gupta, P.K.. *Genetics Classical To Modern*. Rastogi Publications. p. 26. ISBN 9788171338962.

[35] Garrett, Reginald; Grisham, Charles M. (2010). *Biochemistry*. Cengage Learning. p. 535. ISBN 9780495109358.

[36] *New scientist* (IPC Magazines) **152** (2050–2055): 105. 1996.

[37] Castro, Peter; Michael E. Huber (2007). *Marine Biology* (7 ed.). McGraw-Hill. p. 376. ISBN 9780077221249.

[38] Monster fish crushed opposition with strongest bite ever (http://www.smh.com.au/news/science/jaws-of-steel-on-this-fish-tank/2006/11/29/1164777657728.html), smh.com.au

[39] Campbell, Niel A. (1990). *Biology* (2 ed.). Benjamin/Cummings Pub. Co.. p. 560. ISBN 9780805318005.

[40] Richard R. Behringer, Alexander D. Johnson, Robert E. Krumlauf, Michael K. Levine, Nipam Patel, Neelima Sinha, ed (2008). *Emerging model organisms: a laboratory manual, Volume 1* (illustrated ed.). Cold Spring Harbor Laboratory Press. p. 1. ISBN 9780879698720.

[41] Hall, Brian Keith; Benedikt Hallgrímsson, Monroe W. Strickberger (2008). *Strickberger's evolution: the integration of genes, organisms and populations*. Jones & Bartlett Learning. p. 278. ISBN 9780763700669.

[42] Hamilton, Gina. *Kingdoms of Life – Animals (ENHANCED eBook)*. Lorenz Educational Press. p. 9. ISBN 9781429116107.

[43] Maloof, Adam C.; Rose, Catherine V.; Beach, Robert; Samuels, Bradley M.; Calmet, Claire C.; Erwin, Douglas H.; Poirier, Gerald R.; Yao, Nan et al. (17 August 2010). "Possible animal-body fossils in pre-Marinoan limestones from South Australia" (http://www.nature.com/ngeo/journal/vaop/ncurrent/full/ngeo934.html). *Nature Geoscience* **3** (9): 653. Bibcode 2010NatGe...3..653M. doi:10.1038/ngeo934. .

[44] Costa, James T.; Charles Darwin (2009). *The annotated Origin: a facsimile of the first edition of On the origin of species*. Harvard University Press. p. 308. ISBN 9780674032811.

[45] Schopf, J. William (1999). *Evolution!: facts and fallacies*. Academic Press. p. 7. ISBN 9780126288605.

[46] Milsom, Clare; Sue Rigby (2009). *Fossils at a Glance*. John Wiley and Sons. ISBN 9781405193368.

[47] Campbell. Neil A.; Jane B. Reece (2005). *Biology* (7 ed.). Pearson, Benjamin Cummings. p. 526. ISBN 9780805371710.

[48] Seilacher, A., Bose, P.K. and Pflüger, F., A (1998). "Animals More Than 1 Billion Years Ago: Trace Fossil Evidence from India". *Science* **282** (5386): 80–83. Bibcode 1998Sci...282...80S. doi:10.1126/science.282.5386.80. PMID 9756480.

[49] Matz, MV; Frank, TM; Marshall, NJ; Widder, EA; Johnsen, S; Tamara M. Frank, N. Justin Marshall, Edith A. Widder and Sonke Johnsen (2008). "Giant Deep-Sea Protist Produces Bilaterian-like Traces" (http://www.biology.duke.edu/johnsenlab/pdfs/pubs/sea grapes 2008.pdf). *Current Biology* **18** (23): 1–6. doi:10.1016/j.cub.2008.10.028. PMID 19026540. . Retrieved 2008-12-05.

[50] Reilly, Michael (2008-11-20). "Single-celled giant upends early evolution" (http://www.msnbc.msn.com/id/27827279/). MSNBC. . Retrieved 2008-12-05.

[51] Ryan JF, Pang K; NISC Comparative Sequencing Program, Mullikin JC, Martindale MQ, Baxevanis AD (2010) The homeodomain complement of the ctenophore *Mnemiopsis leidyi* suggests that Ctenophora and Porifera diverged prior to the ParaHoxozoa. Evodevo 1(1):9

[52] Bhamrah, H. S.; Kavita Juneja (2003). *An Introduction to Porifera*. Anmol Publications PVT. LTD.. p. 58. ISBN 9788126106752.

[53] Sumich, James L. (2008). *Laboratory and Field Investigations in Marine Life*. Jones & Bartlett Learning. p. 67. ISBN 9780763757304.

[54] Jessop, Nancy Meyer (1970). *Biosphere; a study of life*. Prentice-Hall. p. 428.

[55] Sharma, N. S. (2005). *Continuity And Evolution Of Animals*. Mittal Publications. p. 106. ISBN 9788182930186.

[56] *McGraw-Hill encyclopedia of science & technology: MET-NIC., Volume 11* (http://books.google.com/?id=GS5YAAAAMAAJ) (8 ed.). McGraw-Hill. 1997. p. 59. ISBN 9780079115041. . Retrieved 19 March 2011.

[57] Dunn CW, Hejnol A, Matus DQ, *et al.* (April 2008). "Broad phylogenomic sampling improves resolution of the animal tree of life". *Nature* **452** (7188): 745–9. doi:10.1038/nature06614. PMID 18322464.

[58] Langstroth, Lovell; Libby Langstroth, Todd Newberry, Monterey Bay Aquarium (2000). *A living bay: the underwater world of Monterey Bay*. University of California Press. p. 244. ISBN 9780520221499.

[59] Safra, Jacob E. (2003). *The New Encyclopaedia Britannica, Volume 16*. Encyclopaedia Britannica. p. 523. ISBN 9780852299616.

[60] Kotpal, R. L.. *Modern Text Book of Zoology: Invertebrates*. Rastogi Publications. p. 184. ISBN 9788171339037.

[61] Peters, Kenneth E.; Clifford C. Walters, J. Michael Moldowan (2005). *The Biomarker Guide: Biomarkers and isotopes in petroleum systems and Earth history*. Cambridge University Press. p. 717. ISBN 9780521837620.

[62] Safra, Jacob E. (2003). *The New Encyclopaedia Britannica, Volume 1; Volume 3*. Encyclopaedia Britannica. p. 767. ISBN 9780852299616.

[63] Hyde, Kenneth (2004). *Zoology: An Inside View of Animals*. Kendall Hunt. p. 345. ISBN 9780757509971.

[64] Alcamo, Edward (1998). *Biology Coloring Workbook*. The Princeton Review. p. 220. ISBN 9780679778844.

[65] Holmes, Thom (2008). *The First Vertebrates*. Infobase Publishing. p. 64. ISBN 9780816059584.

[66] Rice, Stanley A. (2007). *Encyclopedia of evolution*. Infobase Publishing. p. 75. ISBN 9780816055159.

[67] Tobin, Allan J.; Jennie Dusheck (2005). *Asking about life*. Cengage Learning. p. 497. ISBN 9780534406530.

[68] Safra, Jacob E. (2003). *The New Encyclopaedia Britannica, Volume 19*. Encyclopaedia Britannica. p. 791. ISBN 9780852299616.

[69] Dawkins, Richard (2005). *The Ancestor's Tale: A Pilgrimage to the Dawn of Evolution*. Houghton Mifflin Harcourt. p. 381. ISBN 9780618619160.

[70] Prewitt, Nancy L.; Larry S. Underwood, William Surver (2003). *BioInquiry: making connections in biology*. John Wiley. p. 289. ISBN 9780471202288.

[71] Schmid-Hempel, Paul (1998). *Parasites in social insects*. Princeton University Press. p. 75. ISBN 9780691059242.

[72] Gilson, Étienne (2004). *El espíritu de la filosofía medieval*. Ediciones Rialp. p. 384. ISBN 9788432134920.

[73] Ruiz-Trillo, I., I; Ruiz-Trillo, Iñaki; Riutort, Marta; Littlewood, D. Timothy J.; Herniou, Elisabeth A.; Baguñà, Jaume (1999). "Acoel Flatworms: Earliest Extant Bilaterian Metazoans, Not Members of Platyhelminthes". *Science* **283** (5409): 1919–1923. Bibcode 1999Sci...283.1919R. doi:10.1126/science.283.5409.1919. PMID 10082465.

[74] Todaro, Antonio. "Gastrotricha: Overview" (http://www.gastrotricha.unimore.it/overview.htm). *Gastrotricha: World Portal*. University of Modena & Reggio Emilia. . Retrieved 2008-01-26.

[75] Kristensen, Reinhardt Møbjerg (2002). "An Introduction to Loricifera, Cycliophora, and Micrognathozoa". *Integrative and Comparative Biology* **42** (3): 641–651. doi:10.1093/icb/42.3.641.

[76] "Biodiversity: Mollusca" (http://web.archive.org/web/20060708083128/http://www.lophelia.org/lophelia/biodiv_6.htm). The Scottish Association for Marine Science. Archived from the original (http://www.lophelia.org/lophelia/biodiv_6.htm) on 2006-07-08. . Retrieved 2007-11-19.

[77] Russell, Bruce J. (Writer), Denning, David (Writer) (2000). *Branches on the Tree of Life: Annelids* (VHS). BioMEDIA ASSOCIATES.

[78] Eernisse, Douglas J., D. J.; Eernisse, Douglas J.; Albert, James S.; Anderson , Frank E. (1 September 1992). "Annelida and Arthropoda are not sister taxa: A phylogenetic analysis of spiralean metazoan morphology". *Systematic Biology* **41** (3): 305–330. doi:10.2307/2992569. JSTOR 2992569.

[79] Eernisse, Douglas J.; Kim, Chang Bae; Moon, Seung Yeo; Gelder, Stuart R.; Kim, Won (1996). "Phylogenetic Relationships of Annelids, Molluscs, and Arthropods Evidenced from Molecules and Morphology". *Journal of Molecular Evolution* (New York: Springer) **43** (3): 207–215. doi:10.1007/PL00006079. PMID 8703086.

[80] Collins, Allen G. (1995). *The Lophophore* (http://www.ucmp.berkeley.edu/glossary/gloss7/lophophore.html). University of California Museum of Paleontology. .

[81] Adoutte, A., A; Adoutte, André; Balavoine, Guillaume; Lartillot, Nicolas; Lespinet, Olivier; Prud'homme, Benjamin; de Rosa, Renaud (2000). "The new animal phylogeny: Reliability and implications". *Proceedings of the National Academy of Sciences* **97** (9): 4453–4456. Bibcode 2000PNAS...97.4453A. doi:10.1073/pnas.97.9.4453. PMC 34321. PMID 10781043.

[82] Passamaneck, Yale J. (2003). "Molecular Phylogenetics of the Metazoan Clade Lophotrochozoa" (http://handle.dtic.mil/100.2/ADA417356) (PDF). p. 124. .

[83] Sundberg, P; Turbeville, JM; Lindh, S; Sundberg, Per; Turbevilleb, J. M.; Lindha, Susanne (2001). "Phylogenetic relationships among higher nemertean (Nemertea) taxa inferred from 18S rDNA sequences". *Molecular Phylogenetics and Evolution* **20** (3): 327–334. doi:10.1006/mpev.2001.0982. PMID 11527461.

[84] Boore, JL; Boore, Jeffrey L.; Staton, Joseph L (2002). "The mitochondrial genome of the Sipunculid Phascolopsis gouldii supports its association with Annelida rather than Mollusca" (http://mbe.oxfordjournals.org/cgi/reprint/19/2/127.pdf) (PDF). *Molecular Biology and Evolution* **19** (2): 127–137. PMID 11801741. . Retrieved 2007-11-19.

[85] Nielsen, Claus (2001). "Bryozoa (Ectoprocta: 'Moss' Animals)" (http://mrw.interscience.wiley.com/emrw/9780470015902/els/article/a0001613/current/abstract). *Encyclopedia of Life Sciences* (John Wiley & Sons, Ltd). doi:10.1038/npg.els.0001613. . Retrieved 2008-01-19.

[86] N.H. Putnam, *et al.* (2007). "Sea anemone genome reveals ancestral eumetazoan gene repertoire and genomic organization". *Science* **317** (5834): 86–94. Bibcode 2007Sci...317...86P. doi:10.1126/science.1139158. PMID 17615350.

[87] Wang, X., X; Wang, Xiujuan; Lavrov Dennis V. (2006-10-27). "Mitochondrial Genome of the homoscleromorph Oscarella carmela ([[Porifera (http://mbe.oxfordjournals.org/cgi/content/abstract/24/2/363)], Demospongiae) Reveals Unexpected Complexity in the

Common Ancestor of Sponges and Other Animals"]. *Molecular Biology and Evolution* **24** (2): 363–373. doi:10.1093/molbev/msl167. PMID 17090697. .

[88] Linnaeus, Carolus (1758) (in Latin). *Systema naturae per regna tria naturae :secundum classes, ordines, genera, species, cum characteribus, differentiis, synonymis, locis.* (http://www.biodiversitylibrary.org/bibliography/542) (10th edition ed.). Holmiae (Laurentii Salvii). . Retrieved September 22, 2008.

Bibliography

- Klaus Nielsen. *Animal Evolution: Interrelationships of the Living Phyla* (2nd edition). Oxford University Press, 2001.
- Knut Schmidt-Nielsen. *Animal Physiology: Adaptation and Environment.* (5th edition). Cambridge University Press, 1997.

External links

- Tree of Life Project (http://tolweb.org/)
- Animal Diversity Web (http://animaldiversity.ummz.umich.edu/site/index.html) – University of Michigan's database of animals, showing taxonomic classification, images, and other information.
- ARKive (http://www.arkive.org) – multimedia database of worldwide endangered/protected species and common species of UK.
- Scientific American Magazine (December 2005 Issue) – Getting a Leg Up on Land (http://www.sciam.com/article.cfm?chanID=sa006&articleID=000DC8B8-EA15-137C-AA1583414B7F0000) About the evolution of four-limbed animals from fish.

ltg:Dzeivinīki koi:Пода rue:Жывы творы

Semotilus_atromaculatus

Common Creek Chub
Scientific classification

Kingdom:	Animalia
Phylum:	Chordata
Class:	Actinopterygii
Order:	Cypriniformes
Family:	Cyprinidae
Genus:	*Semotilus*
Species:	**S. atromaculatus**

Binomial name
Semotilus atromaculatus   (Mitchill, 1818)

Semotilus atromaculatus (also known as the **Common Creek Chub** or simply **"the creek chub"** though this may refer to other *Semotilus* as well) is a small minnow found in the eastern two-thirds of the US and eastern Canada. It inhabits small streams and very small rivers.

Distribution

The creek chub was once abundant in the Lake Erie watershed but is now seen much less frequently. Its current range is the eastern two-thirds of the US and south-east Canada.

Appearance

The creek chub is a small chub that has a greenish-bronze back and a white belly, with a dark stripe running from the nose to the tail along the lateral line. However in some places such as creeks or shallow ditches they have dark brown or black backs, white bellys, and a stripe in the middle that can be brown pink or yellowish. They are usually 5-7 inches in length, but have been known to grow up to 12 inches. They can be identified from other common minnow species by the black moustache on their upper lip, along with a black dot on their dorsal fin. During the breeding season males get small keratin based bumps, called tubercles, on their head which are used in ritualized combat.

Conservation status

The creek chub is considered globally secure and it has been introduced to Utah.

External links

- Information from NatureServe [1]
- Froese, Rainer, and Daniel Pauly, eds. (2008). *"Semotilus atromaculatus"* [2] in FishBase. 02 2008 version.

Fallfish

Fallfish	
Scientific classification	
Kingdom:	Animalia
Phylum:	Chordata
Class:	Actinopterygii
Order:	Cypriniformes
Family:	Cyprinidae
Subfamily:	Leuciscinae
Genus:	*Semotilus*
Species:	**S. corporalis**
Binomial name	
Semotilus corporalis Mitchill, 1817	

The **Fallfish**, *Semotilus corporalis*, is a species in the family Cyprinidae, order Cypriniformes. It is found in the northeastern United States and eastern Canada, where it inhabits streams, rivers, and lake margins. It is fished as a game fish but opinions vary as to its edibility. As far as a game fish goes, it does put up a good fight. The Fallfish tends to be very slimy when held. There is also an audible sucking and gurgling heard when the fish is out of water.

External links

- Froese, Rainer, and Daniel Pauly, eds. (2007). "*Semotilus corporalis*" [1] in FishBase. Apr 2007 version.
- Photo [2]
- Fallfish at Virtual Aquarium [3]

Sandhills_Chub

<table>
<tr><td colspan="2" align="center">Sandhills Chub</td></tr>
<tr><td colspan="2" align="center">Conservation status</td></tr>
<tr><td colspan="2" align="center">Data Deficient (IUCN 2.3)</td></tr>
<tr><td colspan="2" align="center">Scientific classification</td></tr>
<tr><td>Kingdom:</td><td>Animalia</td></tr>
<tr><td>Phylum:</td><td>Chordata</td></tr>
<tr><td>Class:</td><td>Actinopterygii</td></tr>
<tr><td>Order:</td><td>Cypriniformes</td></tr>
<tr><td>Family:</td><td>Cyprinidae</td></tr>
<tr><td>Genus:</td><td>Semotilus</td></tr>
<tr><td>Species:</td><td>S. lumbee</td></tr>
<tr><td colspan="2" align="center">Binomial name</td></tr>
<tr><td colspan="2" align="center">Semotilus lumbee
Snelson & Suttkus, 1978</td></tr>
</table>

The **Sandhills Chub** (*Semotilus lumbee*) is a species of ray-finned fish in the Cyprinidae family. It is found only in the United States.

Source

- Gimenez Dixon, M. 1996. Semotilus lumbee [1]. 2006 IUCN Red List of Threatened Species. [2] Downloaded on 19 July 2007.

Article Sources and Contributors

Semotilus *Source*: http://en.wikipedia.org/w/index.php?title=Semotilus *Contributors*: Amit6, Dawynn, Divingpetrel, Dysmorodrepanis, JR22496, LilHelpa, OhanaUnited, Phn229, Qatter, Waacstats

Cyprinidae *Source*: http://en.wikipedia.org/w/index.php?title=Cyprinidae *Contributors*: Abyssal, Amit6, Anaxial, Apokryltaros, Arthena, Bob1960evens, Bueller 007, Calvero JP, Chris the speller, Chriswaterguy, ClaretAsh, Conversion script, Dawynn, Dennis Brown, Divingpetrel, Dmn, Dolovis, DonaldET3, Droll, Dysmorodrepanis, Eleassar, Erickees, GTBacchus, Gaius Cornelius, Gibsepisg, Gruzd, Gunnar Mikalsen Kvifte, HappyVR, Harry1717, Isaac Read Masters, Isfisk, Josh Grosse, Kerripaul, Knutux, Kurykh, Lerdsuwa, Logan, MPF, Mandarax, Melanochromis, MiltonT, Miwasatoshi, Muhandes, Naddy, Neale Monks, Neil916, Neutrality, Nv8200p, Panellet, Pcb21, Pengo, PenguinJockey, Pharaoh of the Wizards, Quantum bird, Quasirandom, RN1970, Ram-Man, Rettetast, Rich Farmbrough, Richard Barlow, Rickjpelleg, Robina Fox, Ruigeroeland, Sam Hocevar, Shell Kinney, Slavering.dog, Stan Shebs, Syp, T34, Teciltur, Template namespace initialisation script, The Original Fish-Boy, Theo10011, Tide rolls, Tkinias, UkPaolo, Ulrichstill, UtherSRG, Vhorvat, Viridiflavus, Woohookitty, Xandi, Yath, Zangala, 砂翁, 49 anonymous edits

Cypriniformes *Source*: http://en.wikipedia.org/w/index.php?title=Cypriniformes *Contributors*: Abigail-II, Amit6, Anaxial, Ary29, BD2412, Bob the Wikipedian, Conversion script, Dmn, Dora dora, Dysmorodrepanis, Eleassar, Enzino, Frank101, Glenn, GrahamBould, HappyVR, Jauhienij, Josh Grosse, Juansmith, Kerripaul, Kjaergaard, Knutux, Kristof vt, Melanochromis, MiltonT, Neale Monks, Numbo3, Panellet, Pcb21, Pmabee, R'n'B, Ram-Man, Rjwilmsi, Rossami, Scottalter, Stan Shebs, Template namespace initialisation script, Theo10011, Trephination, UtherSRG, Vhorvat, Vina, Vuong Ngan Ha, Wie146, Yath, Ytterzyrkon, 20 anonymous edits

Actinopterygii *Source*: http://en.wikipedia.org/w/index.php?title=Actinopterygii *Contributors*: A2Kafir, AC+79 3888, Abigail-II, Anaxial, Andrew292929, Apokryltaros, Ardric47, AxelBoldt, Bluedustmite, Bob the Wikipedian, Bobblewik, Branka France, Brennenmyers18, CBDunkerson, Cbrodersen, Conversion script, Cyclopia, Daniel, Dawynn, Dj Capricorn, Dmn, Donarreiskoffer, Droll, Duckbill, Dysmorodrepanis, Eliyak, Elonka, Emperorbma, Epbr123, Eric Forste, Evaders99, FunkMonk, GB fan, Gcm, Gdr, GerardM, Gilliam, Ginkgo100, Giraffedata, Glenn, Goudzovski, GrahamBould, Grendelkhan, Gunnar Mikalsen Kvifte, Hadal, HappyVR, Hdahlmo, Helikophis, Höstblomma, Igodard, Infophile, Iph, JLaTondre, Janderk, Jclemens, Jinian, JoJan, Josh Grosse, Juansmith, Kerripaul, Kieff, Kils, Kingpin13, Krellis, Kwamikagami, Logan, Mani1, Mikaey, Mike Cline, Mikko Paananen, MiltonT, Mmernex, Moon&Nature, Muke, Muma, Muriel Gottrop, NatureA16, NeilTarrant, Newone, Ojs, OlEnglish, Omegatron, Paul Pogonyshev, Pcb21, Petter Bøckman, Phil Boswell, PierreAbbat, Pinky sl, Pippu d'Angelo, Plindenbaum, PuzzletChung, Qatter, QueenCake, Radagast3, Ram-Man, Raul654, RetiredWikipedian789, RexNL, Rje, Rjwilmsi, Robertg9, SWAdair, Sandhillcrane, Scottalter, Seglea, Shafei, Sir Lestaty de Lioncourt, Sortior, Srice13, Stan Shebs, Stechlin, Stemonitis, Superxhc, Template namespace initialisation script, TheMathinator, Theo10011, Tkinias, UtherSRG, Vina, Vlmastra, Voyevoda, Vuong Ngan Ha, Whpq, Will Beback, William Avery, Wmfife, Wmleosmith, Wood Thrush, Woohookitty, 91 anonymous edits

Chordate *Source*: http://en.wikipedia.org/w/index.php?title=Chordate *Contributors*: 07joe2k7, 4444hhhh, 61x62x61, A-giau, ACW, Abigail-II, Abrech, AiaiaiPUFFS, Airraid81290, Alansohn, Aldaron, Alien666, An Justified Wikipedian, Andre Engels, Andronimo, Andy Jewell, AngelOfSadness, Angr, Animules, Anna Lincoln, Anomalocaris, Anonymous Dissident, Antandrus, Apokryltaros, Aquapaint, Arcadian, ArchabacteriaNematoda, Arjun01, Arthena, Ashishbhatnagar72, Azhyd, BRG, Baa, Bakerccm, Ballista, Before My Ken, Bendzh, Benjamint444, Betacommand, Betterusername, BigDunc, Bkell, Blanchardb, Bob the Wikipedian, Bobianite, Boccobrock, Boing! said Zebedee, Bomac, Bp41, Brian0918, Bruinfan12, Bubbha, Can't sleep, clown will eat me, CanadianLinuxUser, CanisRufus, CarolinianJeff, Ceolas, Ceph'Ji'Wu, Cephal-odd, Cerealkiller13, Chinasaur, Chordata868, Christakc, Cjakster, Clintp, ColinWhelan, CommonsDelinker, Concordia, Conor12344, Conversion script, Courcelles, Cwmhiraeth, Cyrusc, DARTH SIDIOUS 2, Daniel.levine, Darthgriz98, Dcljr, DennyColt, Deor, DerHexer, Dinoguy2, Dmbaguley, Dogposter, Dougofborg, Dr.K., Dracontes, Dycedarg, Dysepsion, EarthPerson, Ed g2s, Eluchil404, Emperorbma, Enuja, Eog1916, Epbr123, Eric-Wester, Erielhonan, Eug, Eugene van der Pijll, Eve Hall, Evercat, Evil Eccentric, Fenkys, Feydey, Freepsbane, Fujitsum, Funandtrvl, G.A.S, GCarty, Gail, Gdr, Giant Blue Anteater, Glenn, GoldenTorc, Graeme Bartlett, Gravlabs1, GromXXVII, Guacamolesoup, Gurch, Gökhan, Haakon, Hajatvrc, HappyVR, Hapsiainen, Headbomb, HenkvD, Hephaestos, Hydro, Hyenaste, I do not exist, II MusLiM HyBRiD II, Ichigo20, Igodard, Ilovepork, Im.a.lumberjack, Immunize, Iner22, IronGargoyle, Is he back?, Ivan Shmakov, J.delanoy, JForget, JackH, Jackhynes, Jag123, JamesAM, Jameschristopher, Jauhienij, Jimfbleak, JoJan, Joedaguy, Joelmills, Joelr31, JoergenB, Johnrob69, Josh Grosse, Jredytjsfd, Justin, Katya0133, Kbh3rd, Keenan Pepper, Kerotan, Kid45laundry, Kidd Loris, KillerChihuahua, Kingturtle, Klbryant, Kleopatra, Knutux, Korossyl, Kpjas, Kubigula, Kungfuadam, Kurykh, Lawrence Cohen, LeaveSleaves, Lee Daniel Crocker, Lenticel, Lerdsuwa, Ling.Nut, Literacola, Lorenzo Braschi, Lradrama, LuckyWizard, Lyricmac, M Alan Kazlev, M1ss1ontomars2k4, MGC4LIFE, Macrakis, Mad Greg, ManfromButtonwillow, Mani1, Mark t young, Marysunshine, Matijap, Mav, Maximus Rex, Medeis, Megalobingosaurus, Mentifisto, MichaK, Michelleyy, Mild Bill Hiccup, Mintchocolatebear, Miss Madeline, MithrandirAgain, Moon&Nature, Moseley535, Mrthescholz, Msruzicka, Myopic Bookworm, Nabla, Naddy, NatureA16, NawlinWiki, Netoholic, Neutrality, Nickschutz, Nobleeagle, NodnarbLlad, Nwbeeson, Old Father Time, Oleg Alexandrov, Olivier, Orange Suede Sofa, Oxymoron83, PGWG, Pandukht, Pathoschild, Patrick, Paul Pogonyshev, Pcb21, Pdcook, Pegship, Pekaje, Peter Delmonte, Peter Farago, Petter Bøckman, Philcha, Philip Trueman, PierreAbbat, Pippu d'Angelo, Pishogue, Plumbago, Pol430, Postdlf, Prolog, Proxima Centauri, PuzzletChung, Qmwne235, Qwertyus, R9tgokunks, Rich Farmbrough, RichardF, Richardcavell, Rjwilmsi, Rkitko, Rob2001, Robert Foley, Rursus, Samak47, Samsara, Samtheboy, Schneelocke, Scottalter, ScottyBerg, Seaphoto, SebastianHelm, SecWec, Seejyb, Shadowjams, Shafei, ShakingSpirit, SheldonEatWorld, Sheldrake, Shell Kinney, Shrumster, Sidjohn87, Signalhead, Silverxxx, SimonP, Sixpence, Skomorokh, Snigbrook, Snowmanradio, Soliloquial, Some jerk on the Internet, Spinachwrangler, Spotty11222, Stan Shebs, Starry maiden Gazer, Steinsky, Stemonitis, Svanslyck, TUF-KAT, Tablizer, TaborL, Tallcreek, Tbhotch, Template namespace initialisation script, Testicleeze, Tetracube, Thatguyflint, The High Fin Sperm Whale, The Hokkaido Crow, The Master of Mayhem, The Singing Badger, The Thing That Should Not Be, The Winged Yoshi, TheLeopard, Theultimatum, Tide rolls, Timotheus Canens, Tmlotufo, Tpbradbury, Trec'hlid mitonet, Tresiden, Trusilver, Ttiotsw, Twas Now, Twin Bird, Uncle Dick, Unyoyega, UtherSRG, Vandal B, Vanished User 4517, Velella, Vicki Rosenzweig, Vina, Vishyvoice, Vuong Ngan Ha, Wavelength, Williamb, Wiwaxia, Wordsofwisdom101, Wrasse, Xavi1011, Yamazaki-kun, Yath, Yekrats, Yhbestbest, Ynhockey, 594 anonymous edits

Animal *Source*: http://en.wikipedia.org/w/index.php?title=Animal *Contributors*: (jarbarf), -Paul-, 0, 123forever456, 1313 South Harbor Blvd. Anaheim, CA, 14Ave, 168..., 2004-12-29T22:45Z, 28421u2232nfenfcenc, 2D, 4175shelton, 505, 545lljkr, 9258fahsflkh917fas, A-giau, Aaadddaaammm, Aaron Schulz, Aassaas, Abductive, Abigail-II, Abyssal, Academic Challenger, Acalamari, Acdx, Ace ETP, Acebulf, Achiwiki356, Adam78, Adambro, Adashiel, AdjustShift, Adriano1995, Adrigon, Aeonx, Aethon8, Aff123a, Afgheys12, AgainErick, Ahambhavami, Ahoerstemeier, Aidarzver, Aim Here, Airoman, Akendall, Akral, Aksi great, Alainacmitchell, AlanH, Alansohn, Alaska great, Albatross Loves You, Ale jrb, Aleenf1, Alex croff, Alexandria, AlexiusHoratius, Alexy527, Ali, Aliasxerog, Alimon, AllGloryToTheHypnotoad, Ally loves softball, Alterego, Alvin-cs, Amaltheus, Amazins490, Anastrophe, Andjosaus, Andre Engels, Andrewrp, Andux, Andycjp, Angela, Angr, Animum, Ann Stouter, Anna Frodesiak, Anna Lincoln, Anonymous Dissident, Antandrus, Anthere, Anthony717, AnthonyQBachler, Anxietycello, Apostrophe, Apparition11, Aquadiamond3, Aquamarine04, Arakunem, Arcarius, Archer3, Arisa, Art LaPella, Arthur Rubin, Ashmoo, Astropithicus, Aswesee, Atchernev, Athaler, AtikuX, AtonX, Atropos, AuburnPilot, Aude, AutomaticWriting, Avenged Eightfold, Avoided, AxelBoldt, Ayrton Prost, AzaToth, Azhyd, Baccyak4H, Ballista, Bananarofl, Barbary lion, Barneca, Barrylb, Baseballfreakrc, Bassbill, Bassbonerocks, Bbatsell, Bbvslg, Bcasterline, Becie01, Beeblebrox, BeefRendang, Before My Ken, Belligero, Ben-Zin, Ben.c.robers, BenJWoodcroft, Bendzh, Benjamint444, Benjiboi, Bensaccount, Beta m, Bgold, Bidiot, Big Bird, BigChicken, Bigbossfarin, Bigwang6969, Billy775, Biology999, Blanchardb, Blarg3001, Blarrrgy, BlastOButter42, Blaxthos, Blehfu, Bobisbob2, Bobo192, Bobthebuilderbuildsagain, Boccobrock, Boch501, Bomac, BonesBrigade, Bongwarrior, Bookermorgan, Booyabazooka, BorgQueen, Boston, Brambleclawx, Branddobbe, Brian0918, Brianga, Briséis, BrokenSegue, Brossow, Brownsc, BryBla, Bschrader, BuckeyeMoe, Bumbbumb, Bumholio, Bunchofgrapes, Burmeister, CALR, CDepiereux, CJLL Wright, Caesura, Caltas, Can't sleep, clown will eat me, CanadianCaesar, Canderson7, CanisRufus, Capnquackenbush, Carabinieri, CardinalDan, Carl.bunderson, Carlobus, Casper2k3, Causa sui, Cdc, Cdernings, Cekli829, Celithemis, Cephal-odd, Chamal N, Champlax, CharlesHBennett, Chaser, Cherry blossom tree, Chipmunkdavis, Chl, Chmod007, Choanoflagellate, Chocolaterain69lolrofl, Chowchowpoop, Chris j wood, Chris the speller, Chrisd87, Chrislk02, Christian List, Christopher Higgins 93, Chuunen Baka, Ciaccona, Cjc0013, Cleared as filed, Cliff smith, ClockworkSoul, Closedmouth, Cloudy fox 001, Cmorales13990, Cod6user, Cody12343, Coelacan, CohenTheBavarian, Colin Kimbrell, Collieman, Conorsampson, Conti, Conversion script, Cool3, Corncake23, Corpus juris, Corpx, Corvus coronoides, Cory Crowley, Countney, Courcelles, Cow monkey111, Cp111, Cpl Syx, Creeing, Crip killah, Crum375, Crusadeonilliteracy, Crustaceanguy, Crzrussian, CupOfRoses, Cupcakelover101, Curps, Cyclopia, Cyrius, D1sn1gg4, DARTH SIDIOUS 2, DVD R W, Dada43, DagnyB, Dal3164, Dampir, Danger, Daniel C. Boyer, DanielCD, Dar book, DarkFalls, Darkclothes violentpose, Darkmiles22, Darkwind, Darolew, Darth Chyrsaor, Darthgriz98, Dasani, Davewild, David D., David Munch, David0811, Davidlittle9, Dawn Bard, Dcandeto, DeLarge, DeadEyeArrow, Deeptrivia, Delanaschroeder, Delbert Grady, Delirium, Delldot, Demmy, Dendodge, Deor, DerHexer, Dforest, Diagonalfish, Diannaa, DifiCa, Dinno, Dinoguy2, Dinokid, Dipper32, Disavian, Discospinster, Dissimul, DivineAlpha, Dman64, Doc glasgow, Dockcharlotte, Dodo bird, Dojarca, Dolpihnfan123, Donald Albury, Donarreiskoffer, DonnaSG, Dorvaq, Doug Bell, DrMicro, Dracka, Drat, Drc79, Dreaded Walrus, Dreadstar, Drew R. Smith, Drmegabite, Duanedonecker, Dude1818, Dudemeister1234, Dustimagic, Dysepsion, Dysmorodrepanis, Dysprosia, E2eamon, EM85462, EPM, ESkog, EWS23, EamonnPKeane, Earlypsychosis, EarthPerson, East of Borschov, Ebeisher, Ed Poor, Ed g2s, Edk, Edom23, Eightofnine, EikwaR, El C, ElEvErYbOdY, ElePANTS, Eleassar, Elias1richa, Eliezg, Eliz81, Elmo Zat, Eluchil404, Eminem95941, Emperorbma, EncMstr, EncycloPetey, Endangered11, Enigmaman, Enviroboy, Epbr123, Eran, Erick7723, Erik9, Esanchez7587, Escape Orbit, EscapingLife, Esn, Esurnir, Eu.stefan, Eugene van der Pijll, Everyking, Excirial, Exert, Eyu100, FF2010, Fabiform, Facts707, Fang 23, Farrahi, Farsight001, Feedloadr, Felizdenovo, Femto, Ferel destroyer, Fgr1994, Fieldday-sunday, Fil21, Filelakeshoe, Finlay McWalter, FireOfTheNight, Firsfron, FisherQueen, Flat turds, Flewis, Flo98, Flowerpotman, Foosher, Francisco Valverde, Frap, Freakofnurture, Fredvanner, FreplySpang, Fritoflyer, Frubey, Fullz, Furkaocean, Furkhaocean, Fvw, Fæ, GB fan, GTPoompt, Gaaras-brother, Gabbe, Gadfium, Gaff, Gaius Cornelius, Gakrivas, Galoubet, Gamesmaster9, Gavrilis, Gcjdavid, Gdr, Geni, GeorgeMoney, Gfoley4, Gianfranco, Giftlite, Gilliam, Gimmetrow, Gingerman1234567, Ginkgo100, Gjd001, Glenn, Gnostrat, Go for it!, Gogo Dodo, Goldom, Goodnightmush, Gothicrocka, GraemeL, Graham87, GrahamBould, Grblundell, Greeener, Greswik, Grundle2600, Gscshoyru, Guanaco, Gundato, Gurch, Gutworth, Gwernol, Gypsy Trash, H3110kitty, Hadal, Haham hanuka, HalJor, HalfShadow, Halfcrowd123, Halo master chief, Halosaver, Ham1999, HappyBlades, HappyKarMar, Happyface0000, HarlandQPitt, Hbomb194, Heebiejeebieclub, Heegoop, Heimstern, Helicoptor, Hellbus, Hello32020, Heman, Hephaestos, Hermanjr, Hibana, Hihi582, Hordaland, Hotie1234, Hu12, HughesJohn, HunterAmor, Hut 8.5, Hyebeh, I dream of horses, IAMTHEEGGMAN, ICAPTCHA, II MusLiM HyBRiD II, Ianml, Icestorm815, IdahoBoy, Idont Havaname, Ifnord, Indoles, Iner22, Insanity Incarnate, Intelligentsium, Into The Fray, Irbisgreif, Iridescent, Ixfd64, J. Spencer, J.delanoy, JAKLAM, JForget, JIP, JLaTondre, JNW, JTSomers, JULIA RADESCU IS A SKIVER!, JYolkowski, JamesAM, Jamesofur, Jamsamree, Janderk, Jannex, Jasongetsdown, Jauerback, Javert, Jawsx188, Jay, Jay2332, JayW, Jaymckenzie, JedOs, Jeendan, Jefffire, Jeffrey Mall, Jeffrey O. Gustafson,

Image Sources, Licenses and Contributors

file:TincaTincaWeerribben.JPG *Source*: http://en.wikipedia.org/w/index.php?title=File:TincaTincaWeerribben.JPG *License*: unknown *Contributors*: Piet Spaans

Image:Giant Barb.jpg *Source*: http://en.wikipedia.org/w/index.php?title=File:Giant_Barb.jpg *License*: unknown *Contributors*: User:Lerdsuwa

Image:Carpe miroir de 17kg.jpg *Source*: http://en.wikipedia.org/w/index.php?title=File:Carpe_miroir_de_17kg.jpg *License*: unknown *Contributors*: Lolo770461, Vmenkov, 1 anonymous edits

Image:GoldfishPearl.jpg *Source*: http://en.wikipedia.org/w/index.php?title=File:GoldfishPearl.jpg *License*: unknown *Contributors*: User:Fanghong

Image:Acheilognathus longipinnis2.jpg *Source*: http://en.wikipedia.org/w/index.php?title=File:Acheilognathus_longipinnis2.jpg *License*: unknown *Contributors*: KENPEI

Image:Danio kerri.jpg *Source*: http://en.wikipedia.org/w/index.php?title=File:Danio_kerri.jpg *License*: unknown *Contributors*: User:Lerdsuwa

Image:Pseudogobio esocinus(Hamamatsu,Shizuoka,Japan).jpg *Source*: http://en.wikipedia.org/w/index.php?title=File:Pseudogobio_esocinus(Hamamatsu,Shizuoka,Japan).jpg *License*: unknown *Contributors*: User:Seotaro

Image:Hypophthalmichthys molitrix adult.jpg *Source*: http://en.wikipedia.org/w/index.php?title=File:Hypophthalmichthys_molitrix_adult.jpg *License*: unknown *Contributors*: user:Tdk

Image:Labeo rohita.JPG *Source*: http://en.wikipedia.org/w/index.php?title=File:Labeo_rohita.JPG *License*: unknown *Contributors*: User:Khalid Mahmood

Image:Hemitremia flammea.jpg *Source*: http://en.wikipedia.org/w/index.php?title=File:Hemitremia_flammea.jpg *License*: unknown *Contributors*: Haplochromis, Liné1

File:Aland.jpg *Source*: http://en.wikipedia.org/w/index.php?title=File:Aland.jpg *License*: unknown *Contributors*: Kristof vt

Image:Notropis hypselopterus.jpg *Source*: http://en.wikipedia.org/w/index.php?title=File:Notropis_hypselopterus.jpg *License*: unknown *Contributors*: Eugene van der Pijll, Haplochromis, IP 84.5, Liné1

Image:Phoxinus oxycephalus jouyi1.jpg *Source*: http://en.wikipedia.org/w/index.php?title=File:Phoxinus_oxycephalus_jouyi1.jpg *License*: unknown *Contributors*: KENPEI

File:Rutilus rubilio.jpg *Source*: http://en.wikipedia.org/w/index.php?title=File:Rutilus_rubilio.jpg *License*: unknown *Contributors*: User:Etrusko25

Image:Trigonostigma somphongsi.jpg *Source*: http://en.wikipedia.org/w/index.php?title=File:Trigonostigma_somphongsi.jpg *License*: unknown *Contributors*: User:Lerdsuwa

Image:Mylopharyngodon piceus.jpg *Source*: http://en.wikipedia.org/w/index.php?title=File:Mylopharyngodon_piceus.jpg *License*: unknown *Contributors*: Leo Nico, (United States Geological Survey) (Original uploader was Zp at cs.wikipedia)

Image:Hemigrammocypris rasborella(Fujieda-shi,Shizuoka-ken,Japan).jpg *Source*: http://en.wikipedia.org/w/index.php?title=File:Hemigrammocypris_rasborella(Fujieda-shi,Shizuoka-ken,Japan).jpg *License*: unknown *Contributors*: User:Seotaro

Image:Steinbeisser 001.jpg *Source*: http://en.wikipedia.org/w/index.php?title=File:Steinbeisser_001.jpg *License*: unknown *Contributors*: J.C. Harf (User Wiki-Harfus on de.wikipedia)

Image:Nemac fasci 080519 9380 ckoep.jpg *Source*: http://en.wikipedia.org/w/index.php?title=File:Nemac_fasci_080519_9380_ckoep.jpg *License*: unknown *Contributors*: User:Wie146

Image:Chinese algae eater.jpg *Source*: http://en.wikipedia.org/w/index.php?title=File:Chinese_algae_eater.jpg *License*: unknown *Contributors*: User:Pseudogastromyzon

Image:Erimyzon sucetta.jpg *Source*: http://en.wikipedia.org/w/index.php?title=File:Erimyzon_sucetta.jpg *License*: unknown *Contributors*: Haplochromis, Liné1

Image:Thicktail Chub.jpg *Source*: http://en.wikipedia.org/w/index.php?title=File:Thicktail_Chub.jpg *License*: unknown *Contributors*: California Department of Fish and Game

Image:Epalzeorhynchos bicolor.jpg *Source*: http://en.wikipedia.org/w/index.php?title=File:Epalzeorhynchos_bicolor.jpg *License*: unknown *Contributors*: Original uploader was Caveman99 at de.wikipedia

Image:Hypsospondylus.JPG *Source*: http://en.wikipedia.org/w/index.php?title=File:Hypsospondylus.JPG *License*: unknown *Contributors*: User:Ghedoghedo

file:Pristella maxillaris.jpg *Source*: http://en.wikipedia.org/w/index.php?title=File:Pristella_maxillaris.jpg *License*: unknown *Contributors*: User:Makro Freak

File:Red Pencil Icon.png *Source*: http://en.wikipedia.org/w/index.php?title=File:Red_Pencil_Icon.png *License*: unknown *Contributors*: User:Peter coxhead

Image:BranchiostomaLanceolatum PioM.svg *Source*: http://en.wikipedia.org/w/index.php?title=File:BranchiostomaLanceolatum_PioM.svg *License*: unknown *Contributors*: Piotr Michał Jaworski; PioM EN DE PL

Image:Pacific hagfish Myxine.jpg *Source*: http://en.wikipedia.org/w/index.php?title=File:Pacific_hagfish_Myxine.jpg *License*: unknown *Contributors*: EugeneZelenko, Infrogmation, Liné1, Lmozero, Snek01, Stan Shebs, 1 anonymous edits

Image:Branchiostoma lanceolatum.jpg *Source*: http://en.wikipedia.org/w/index.php?title=File:Branchiostoma_lanceolatum.jpg *License*: unknown *Contributors*: User:Biopics

Image:BU Bio.jpg *Source*: http://en.wikipedia.org/w/index.php?title=File:BU_Bio.jpg *License*: unknown *Contributors*: Original uploader was Elapied at fr.wikipedia

Image:Balanoglossus 01.png *Source*: http://en.wikipedia.org/w/index.php?title=File:Balanoglossus_01.png *License*: unknown *Contributors*: Philcha (talk) Original uploader was Philcha at en.wikipedia

Image:Sandstar 300.jpg *Source*: http://en.wikipedia.org/w/index.php?title=File:Sandstar_300.jpg *License*: unknown *Contributors*: Eugene van der Pijll, Haplochromis, JoJan, Liné1, Lucero del Alba, Lycaon, Xhienne

Image:Haikouichthys4.png *Source*: http://en.wikipedia.org/w/index.php?title=File:Haikouichthys4.png *License*: unknown *Contributors*: User:Giant Blue Anteater

file:Animal diversity.png *Source*: http://en.wikipedia.org/w/index.php?title=File:Animal_diversity.png *License*: unknown *Contributors*: User:Anilocra, User:Bkmiles, User:Kevincollins123, User:Lviatour, User:Medeis, User:Nhobgood, User:Opoterser, User:Panda3, user:Australianplankton

File:Mitosis-flourescent.jpg *Source*: http://en.wikipedia.org/w/index.php?title=File:Mitosis-flourescent.jpg *License*: unknown *Contributors*: -

File:Dunkleosteus BW.jpg *Source*: http://en.wikipedia.org/w/index.php?title=File:Dunkleosteus_BW.jpg *License*: unknown *Contributors*: User:ArthurWeasley

File:Vernanimalcula.jpg *Source*: http://en.wikipedia.org/w/index.php?title=File:Vernanimalcula.jpg *License*: unknown *Contributors*: Ricky81682, Smith609

File:AnimalsRelativeNumbers.png *Source*: http://en.wikipedia.org/w/index.php?title=File:AnimalsRelativeNumbers.png *License*: unknown *Contributors*: Nick Beeson (Nwbeeson at en.wikipedia)

File:Elephant-ear-sponge.jpg *Source*: http://en.wikipedia.org/w/index.php?title=File:Elephant-ear-sponge.jpg *License*: unknown *Contributors*: Elapied, Liné1, Maksim

File:Superbfairywrenscropped.jpeg *Source*: http://en.wikipedia.org/w/index.php?title=File:Superbfairywrenscropped.jpeg *License*: unknown *Contributors*: Benjamint444, CommonsDelinker, Pengo, Smihael, Tony Wills, Wst, 1 anonymous edits

File:Sympetrum flaveolum - side (aka).jpg *Source*: http://en.wikipedia.org/w/index.php?title=File:Sympetrum_flaveolum_-_side_(aka).jpg *License*: unknown *Contributors*: user:Aka

File:Bedford's Flatworm.jpg *Source*: http://en.wikipedia.org/w/index.php?title=File:Bedford's_Flatworm.jpg *License*: unknown *Contributors*: Jan Derk

File:Grapevinesnail 01.jpg *Source*: http://en.wikipedia.org/w/index.php?title=File:Grapevinesnail_01.jpg *License*: unknown *Contributors*: Jürgen Schoner

File:Carolus Linnaeus (cleaned up version).jpg *Source*: http://en.wikipedia.org/w/index.php?title=File:Carolus_Linnaeus_(cleaned_up_version).jpg *License*: unknown *Contributors*: Original painting by Alexander Roslin. Digitally improved by Greg L.

file:Creek Chub, Semotilus atromaculatus.jpg *Source*: http://en.wikipedia.org/w/index.php?title=File:Creek_Chub,_Semotilus_atromaculatus.jpg *License*: unknown *Contributors*: Brian Gratwicke

file:Status iucn2.3 blank.svg *Source*: http://en.wikipedia.org/w/index.php?title=File:Status_iucn2.3_blank.svg *License*: unknown *Contributors*: Pengo

Printed by Books on Demand GmbH, Norderstedt / Germany